AF818211

Bio-optical Properties of Indian Coastal Waters

Sourav Das
Editor

Bio-optical Properties of Indian Coastal Waters

Springer

Editor
Sourav Das
Department of Chemistry
Brainware University
Kolkata, West Bengal, India

ISBN 978-3-032-08713-3 ISBN 978-3-032-08714-0 (eBook)
https://doi.org/10.1007/978-3-032-08714-0

© The Editor(s) (if applicable) and The Author(s), under exclusive license to Springer Nature Switzerland AG 2026

This work is subject to copyright. All rights are solely and exclusively licensed by the Publisher, whether the whole or part of the material is concerned, specifically the rights of translation, reprinting, reuse of illustrations, recitation, broadcasting, reproduction on microfilms or in any other physical way, and transmission or information storage and retrieval, electronic adaptation, computer software, or by similar or dissimilar methodology now known or hereafter developed.
The use of general descriptive names, registered names, trademarks, service marks, etc. in this publication does not imply, even in the absence of a specific statement, that such names are exempt from the relevant protective laws and regulations and therefore free for general use.
The publisher, the authors and the editors are safe to assume that the advice and information in this book are believed to be true and accurate at the date of publication. Neither the publisher nor the authors or the editors give a warranty, expressed or implied, with respect to the material contained herein or for any errors or omissions that may have been made. The publisher remains neutral with regard to jurisdictional claims in published maps and institutional affiliations.

This Springer imprint is published by the registered company Springer Nature Switzerland AG
The registered company address is: Gewerbestrasse 11, 6330 Cham, Switzerland

If disposing of this product, please recycle the paper.

Dedicated to my PhD guide Prof. Sugata Hazra, Ex-Director, School of Oceanographic Studies, Jadavpur University, Kolkata, India.

Preface

Currently, Estuarine Water Optics, Coastal Water Science, or Estuarine Science are very emerging in the state and yet to be established disciplines in the world of science. Students from various backgrounds are studying such courses during their undergraduate or graduate programs and carrying out their research in these types of specialized sub-discipline. Ocean optics is a very established research sector in the present world, but the estuarine optical study is much more difficult and critical, and it is very much site- or location-specific. Therefore, ocean optics laboratories are all trying to solve such difficulties by producing different bio-optical algorithms. Under these types of algorithms, researchers are trying to study ocean colour or coastal colour.

Ocean optics is the study of how light interacts with water and the materials in water. Although research often focuses on the sea, the field broadly includes rivers, lakes, inland waters, coastal waters, and large ocean basins. How light acts in water is critical to how ecosystems function underwater. The color of the ocean is determined by the interaction of sunlight with substances or particles present in seawater (such as sediments) and chlorophyll, a green pigment found in most phytoplankton species.

Ocean colour analysis has proven to be an efficient tool for determining the health of the ocean by measuring oceanic biological activity through optical means. Phytoplankton pigment acts as an indicator of the equilibrium of CO_2 concentration between the atmosphere and the ocean. The other aquatic optical components, such as chromophoric dissolved organic matter (CDOM) and total suspended matter (TSM), also act as ecological indicators. The oceanic colour change due to chlorophyll_a provides the level of primary productivity, whereas the colour change due to CDOM can be attributed to the level of pollution. Higher particulate matter in suspension can be directly linked to wave and current energy. The advent of satellite technology has given a substantial boost to ocean colour research. The ability of optical sensors to map the spatial and temporal patterns of ocean colour over regional and global scales has provided important insights into the fundamental bio-optical properties and bio-physical processes occurring in the aquatic media. However, the satellite-based retrieval of ocean colour parameters demands an accurate bio-optical algorithm, especially in coastal waters, and such algorithms require a large in situ database for a specific area.

To address these issues along the Indian coastline, the Satellite Coastal and Oceanographic Research (SATCORE) programme was initiated by the Indian Government. The Indian National Centre for Ocean Information Services (INCOIS), an autonomous body under the Ministry of Earth Sciences, Govt. of India, started coordinating the SATCORE programme from the XI Five-Year Plan with the aim of long-term measurements of bio-optical properties in the Indian coastal waters. The aim was to validate the satellite-derived ocean products in coastal waters and their subsequent application in operational services, such as the location of potential fishing zones (PFZs) and the detection and monitoring of algal blooms (ABs). Apart from these, the other objectives of the SATCORE programme are to develop new algorithms for ocean colour satellite sensors, improve atmospheric correction schemes, and apply ocean colour data for modelling primary productivity. SATCORE also supplements the global initiative, Chlorophyll Global Integration Network (ChloroGIN), which aims towards the generation of ocean colour data products from operational satellite sensors at near real-time (NRT) for operational applications. A valuable dataset of physical, optical, and bio-geo-chemical parameters was generated from the SATCORE programme for the period 2008–2013. The measured parameters include: Chlorophyll_a (Chl-a), coloured dissolved organic matter (CDOM), total suspended matter (TSM), inherent optical properties (IOP), including absorption and scattering coefficients, apparent optical properties (AOP) measured using hyperspectral radiometer, measurement using a conductivity-temperature-depth (CTD) instrument, nutrients, aerosol optical thickness (AOT), and meteorological parameters (Met) measured using an Autonomous Weather Station (AWS).

A substantial number of research articles have been published in reputed journals addressing the bio-optical properties of Indian coastal waters. Despite this, the coastal waters of India are highly diverse, with each region displaying unique optical and biogeochemical characteristics. Most published studies focus on specific locations, resulting in findings that are limited in geographic scope and not representative of the broader Indian coastline. Recognizing this gap, the present edited volume brings together a comprehensive overview of the bio-optical characteristics of various Indian coastal regions. This is the first consolidated effort to document and present such information in a single book, offering a unified understanding of the subject. Although several high-quality books exist in the field of aquatic sciences, very few concentrate on recent developments in bio-optical parameters or modelling techniques, particularly in coastal environments. Importantly, none has specifically targeted the Indian coastal context. *Bio-optical Properties of Coastal Waters* introduces state-of-the-art methodologies, current research trends, and future perspectives in the bio-optical modelling of optically active components in Indian coastal waters. Emphasis is placed on the practical applications of these models, particularly for water quality monitoring using satellite-based remote sensing data and field measurements, rather than focusing solely on theoretical aspects of optical radiometry.

This book not only covers applications but also discusses new possibilities for making the bio-optical theories operational, a concept that is of great interest to both government and private sector organizations. In addition, it addresses not only the physical theory that makes bio-optical modelling possible but also the implementation and applications of bio-optical modelling in inland waters.

Early chapters introduce the concepts of bio-optical modelling, the classification of bio-optical models, and existing and developing satellite capabilities. Later chapters target specific optically active components (OACs) for inland waters and present the current status and future direction of bio-optical modelling for these OACs. Concluding sections provide an overview of a governance strategy for Indian inland water monitoring based on Earth observation and bio-optical modelling and also explain the progress in Ocean Colour Remote Sensing (OCRS) through satellite missions, research infrastructure, and capacity-building programs.

The editor and authors will be extremely grateful to those who make substantive suggestions or point out unintentional errors.

Kolkata, West Bengal, India

Sourav Das

Acknowledgments

I would like to express my sincere gratitude to Dr. Sanjiba Kumar Baliar Singh (Scientist, INCOIS, India) and Dr. Sudarsan Rao Pandi (Project Scientist, NCPOR, Goa, India) for their valuable assistance in reviewing several chapters of this book. I am also thankful to Mr. Aaron Schiller at Springer Nature for his kind support and encouragement. My heartfelt thanks go to the Vice Chancellor of Brainware University (BWU), Kolkata, for his continued support and encouragement throughout this book project. Finally, I extend my special thanks to my wife, Mrs. Jaya Das, for her help in editing various tables, photos, and images.

Contents

About the Editor

Sourav Das Dr. Sourav Das is currently an Assistant Professor in the Department of Chemistry at Brainware University, Kolkata, India (email id: sourav.biooptics@gmail.com). He completed his doctoral degree (PhD) from the School of Oceanographic Studies at Jadavpur University, India. Moreover, he was a post-doctoral fellow for the Living Delta Hub at Jadavpur University, Kolkata. His research interests include environmental pollution, biogeochemistry of marine and coastal water, ocean remote sensing, optical properties of water (inherent and apparent optical properties), and air quality monitoring. He has more than fifteen (15) years of research and teaching experience and has published more than 66 research articles in reputed peer-reviewed journals, including three (03) edited books and three (03) authored books. Moreover, he has contributed more than thirty-two (32) peer-reviewed book chapters. He has more than 850 citations and an h-index of 16 (www.researchgate.net/profile/Sourav-Das-34). He is an eminent Environmental Scientist in India .

Chapter 1
Introduction to Bio-Optical Properties of Coastal Waters Special Emphasis on the Indian Coast

Sourav Das

Abstract This book offers a comprehensive exploration of bio-optical variability and ocean color remote sensing in India's coastal waters. It highlights the importance of satellite-derived chlorophyll-a for monitoring ecosystem health, detecting algal blooms, and supporting fisheries, emphasizing the need for region-specific calibration. The complex optical properties shaped by freshwater influx, river discharges, and seasonal monsoons are examined, with a focus on differences between the Arabian Sea and the Bay of Bengal. The role of phytoplankton dynamics, CDOM variability, and particle size structure is analyzed to understand ecological processes and human impacts. Case studies from diverse coastal environments, including estuaries and river plumes, illustrate the use of remote sensing in tracking productivity, pollution, and water quality. The progress in Ocean Colour Remote Sensing in India, supported by training and institutional efforts, is also discussed. This work aims to advance sustainable management and scientific understanding of India's diverse and dynamic coastal ecosystems.

Keywords Bio-optical variability · Ocean color remote sensing · Chlorophyll-a · Coastal ecosystems · Chromophoric dissolved organic matter (CDOM) · Indian coastline

1 An Overview

Coastal waters are dynamic environments characterized by significant variability in their biological, physical, and chemical properties. These regions act as important interfaces between land and sea, serving as major sources of both dissolved and particulate matter that are transported to the open ocean. Due to their high productivity, a considerable amount of organic material is produced in the surface waters of coastal zones, eventually sinks, and accumulates in the underlying sediments.

S. Das (✉)
Department of Chemistry, Brainware University, Kolkata, West Bengal, India

© The Author(s), under exclusive license to Springer Nature Switzerland AG 2026

S. Das (ed.), *Bio-optical Properties of Indian Coastal Waters*,
https://doi.org/10.1007/978-3-032-08714-0_1

Along coastal margins, especially, this organic matter may become permanently buried, playing a key role in long-term carbon sequestration (Hedges & Keil, 1995).

The complexity of coastal systems is further heightened in areas influenced by riverine inputs. These zones receive substantial volumes of freshwater that vary seasonally and are often rich in suspended sediments, organic compounds, and nutrients. Such variability introduces significant spatial and temporal heterogeneity, making it difficult to accurately quantify the role of coastal regions in the global biogeochemical carbon cycle. This uncertainty is particularly evident in the estimation of coastal carbon fluxes and their contribution to broader carbon dynamics (Wollast, 1991).

In this context, satellite-based remote sensing has emerged as a powerful tool for observing and interpreting coastal processes. Its ability to deliver frequent, large-scale, and synoptic measurements makes it especially useful for tracking dynamic phenomena in coastal waters. The development of a variety of bio-optical algorithms has further enhanced the applicability of ocean color remote sensing. These algorithms enable the retrieval of important seawater constituents, such as phytoplankton biomass and primary productivity—parameters that are critical for evaluating carbon cycling in marine ecosystems (Platt et al., 1988; Muller-Karger et al., 1991).

Moreover, ocean color data have been extensively utilized to assess key indicators of water quality, including water transparency, concentrations of suspended sediments, and the occurrence of harmful algal blooms (Carder & Steward, 1985; Tassan, 1994; Kahru & Mitchell, 1998; Stumpf et al., 1999). Despite these advancements, the inherent optical complexity of coastal waters, resulting from the interplay of various constituents and environmental factors, continues to present challenges for accurately interpreting satellite-derived data. Addressing these challenges remains essential for improving our understanding and monitoring of coastal ecosystems.

Bio-optical properties—broadly defined as the relationships between the inherent and apparent optical characteristics of seawater and its biological constituents—serve as a foundational element of modern ocean optics. These properties govern how sunlight is absorbed, scattered, and transmitted through the water column, thereby influencing a range of critical processes, including primary productivity, heat distribution, and the effectiveness of remote sensing techniques.

In optically complex coastal regions, such as those surrounding India, the study of bio-optical interactions becomes both especially challenging and vitally important. These areas are marked by high concentrations of suspended sediments, significant levels of colored dissolved organic matter (CDOM), and considerable variability in phytoplankton populations—all of which contribute to complex and dynamic optical conditions. Understanding and accurately characterizing these interactions is essential for improving our ability to monitor and model coastal ecosystems using satellite-based and in situ observational tools.

Oceanic waters are generally classified into two main categories—Case 1 and Case 2—based on the characteristics and variability of their optical properties, a framework first proposed by Morel and Prieur (1977) and later refined by Gordon and Morel (1983). This classification helps in understanding and modeling how light interacts with substances present in different types of marine environments, which is critical for interpreting remote sensing data and bio-optical measurements.

Case 1 waters are typically found in the open ocean, where optical properties are largely governed by phytoplankton and the substances directly associated with them. These include colored dissolved organic matter (CDOM) and detrital particles, both of which tend to co-vary with chlorophyll concentration. In these waters, chlorophyll is often used as a proxy for estimating other optically active constituents, making them relatively easier to model and analyze using standard ocean color remote sensing algorithms.

On the other hand, Case 2 waters are found predominantly in coastal and inland regions, where the optical environment is much more complex. In these waters, light absorption and scattering are influenced by a wider variety of substances, such as suspended sediments, CDOM, and inorganic particles, which may originate from terrestrial runoff, riverine discharge, or bottom resuspension. Unlike in Case 1 waters, these constituents in Case 2 waters do not necessarily correlate with phytoplankton concentration or productivity, making remote sensing and optical modeling more challenging.

Due to their high variability and the diverse sources of optically active materials, coastal and inland waters are generally categorized as Case 2 waters (Sathyendranath, 2000). Accurately characterizing these environments requires advanced algorithms and region-specific calibration to account for the complex interplay of biological and non-biological constituents affecting light penetration and reflectance.

Historically, ocean color remote sensing research has predominantly focused on Case 1 waters—open ocean regions where phytoplankton and their associated by-products, such as colored dissolved organic matter (CDOM) and detritus, are the primary contributors to optical variability. In these relatively homogeneous environments, standard ocean color algorithms have shown a reasonable degree of accuracy in estimating chlorophyll-a concentrations, which serve as a proxy for phytoplankton biomass and primary productivity.

In contrast, Case 2 waters—typically found in coastal, estuarine, and inland regions—present a far more complex optical environment. These waters contain a diverse mix of optically active substances, including suspended sediments, CDOM, and various inorganic and organic particulates, which often vary independently of phytoplankton. As a result, conventional algorithms developed for Case 1 waters often perform poorly in Case 2 environments, leading to significant errors in the estimation of biogeochemical parameters (Das et al., 2017; Darecki & Stramski, 2004).

One illustrative example of this challenge comes from studies of the Orinoco River plume. In this region, Hochman et al. (1994) found that up to 50% of the chlorophyll-a concentration estimated using data from the Coastal Zone Color Scanner (CZCS) may have been an artifact, largely due to the presence of elevated levels of CDOM. This finding highlights the limitations of standard remote sensing approaches in optically complex waters and underscores the need for specialized, region-specific algorithms that can better differentiate between various constituents affecting light absorption and scattering.

These challenges emphasize the importance of developing advanced bio-optical models and conducting targeted field validation campaigns to improve the accuracy and reliability of satellite-derived products in Case 2 waters. Addressing these issues is essential for expanding the applicability of remote sensing technologies to coastal zones, which are ecologically and economically vital yet remain underrepresented in global ocean color research.

This issue is not isolated. In the northern Gulf of Mexico, Carder et al. (1989) documented large variations in the CDOM-to-chlorophyll ratio, a factor that complicates both empirical and semi-analytical approaches to interpreting ocean color data. In many river-influenced coastal systems, the interplay among physical, chemical, and biological processes further obscures optical clarity, and the lack of comprehensive knowledge regarding local optical properties continues to hinder the reliable application of remote sensing tools in these regions.

However, India's extensive coastline, stretching over 7500 km, encompasses a diverse array of marine ecosystems, including estuaries, mangroves, coral reefs, and continental shelf zones. These coastal waters are integral to the biogeochemical and ecological processes of the northern Indian Ocean, supporting rich biodiversity and providing livelihoods to millions. Understanding the optical properties of these waters, particularly the interaction of light with biological and non-biological constituents, is critical for advancing marine research, resource management, and environmental monitoring.

2 Significance of Bio-Optical Studies in Coastal Waters

India's coastal zones experience significant anthropogenic pressures resulting from a range of human activities, including riverine discharges, agricultural runoff, urban wastewater, and industrial effluents. These inputs introduce a complex mix of pollutants and nutrients into the marine environment, substantially altering the optical characteristics of the water column. Such changes influence the way light penetrates and interacts with seawater, affecting primary productivity, species composition, and overall ecological balance in these sensitive regions.

In addition to these anthropogenic influences, natural seasonal variability—primarily driven by the monsoon cycle—further complicates the bio-optical landscape. The monsoon brings about pronounced fluctuations in salinity, temperature, and nutrient availability due to increased freshwater influx, sediment load, and upwelling processes. These dynamic physical and chemical shifts create highly variable conditions that are both a challenge and an opportunity for bio-optical research.

Effectively monitoring and modeling these changes requires a deep understanding of the relationships between optical signals and in situ biological variables. Key parameters such as chlorophyll-a concentration, phytoplankton functional types, and particle size distributions must be accurately quantified to interpret optical data correctly. This is where bio-optical research plays a critical role. It provides the scientific foundation for diagnosing water quality, tracking ecological shifts, and understanding biogeochemical processes.

Moreover, bio-optical studies are essential for the calibration and validation of satellite-based ocean color sensors. These sensors rely on precise bio-optical algorithms to detect large-scale spatial and temporal patterns in biological productivity, suspended matter, and environmental health. As such, robust bio-optical datasets from India's coastal waters are crucial for enhancing the accuracy and reliability of remote sensing applications, ultimately supporting better marine resource management, environmental monitoring, and policy-making.

3 Components of Bio-Optical Properties

Bio-optical properties are typically classified into two main categories:

3.1 Inherent Optical Properties (IOPs)

These are properties of water that depend only on the medium and not on the ambient light field. IOPs include:

- Absorption coefficient (a)**:** Measures how much light is absorbed per unit distance.
- Scattering coefficient (b)**:** Measures how much light is scattered per unit distance.
- Backscattering coefficient (bb)**:** Measures the portion of scattered light that is redirected in the backward direction.

In coastal waters, these are primarily influenced by phytoplankton pigments, CDOM, and suspended particulate matter.

3.2 Apparent Optical Properties (AOPs)

These are influenced by both the IOPs and the geometric structure of the ambient light field. Examples include:

- **Diffuse attenuation coefficient (Kd)**: Describes the rate at which light diminishes with depth.
- **Remote sensing reflectance (Rrs)**: The ratio of water-leaving radiance to downwelling irradiance just above the surface.

AOPs are crucial for interpreting satellite-based ocean color data.

Understanding the linkages between IOPs and AOPs, and how these relate to biological variables, is central to bio-optical modeling and remote sensing applications.

4 Challenges in Indian Coastal Waters

In contrast to open ocean waters (Case I), which are primarily influenced by phytoplankton and their by-products, coastal waters (Case II)—such as those surrounding India—are affected by a much broader and more complex range of substances. These influences include:

- Terrestrial runoff, which introduces large amounts of sediments and organic matter from land.
- Tidal mixing, which resuspends bottom sediments, alters light penetration and water clarity.
- Major river discharges, such as those from the Ganges, Brahmaputra, Godavari, and Mahanadi, which contribute freshwater, sediments, and nutrients to coastal zones.
- Eutrophication and algal blooms, driven by excessive nutrient inputs from agricultural and urban sources.
- Monsoonal and interannual variability significantly affect salinity, temperature, turbidity, and overall primary productivity.

These complex and dynamic conditions make the application of conventional optical models—typically developed for simpler Case I waters—largely inadequate for Indian coastal environments. As a result, there is a critical need for the development of region-specific bio-optical algorithms that account for the unique physical, chemical, and biological characteristics of the Indian subcontinent's diverse coastal waters. Such tailored approaches are essential for improving the accuracy of remote sensing observations and for advancing the understanding and management of these ecologically vital regions.

5 Advances and Applications

Recent decades have witnessed significant advancements in in situ measurements, hyperspectral sensors, and radiative transfer modeling. This has enabled the development of robust empirical and semi-analytical models linking optical signals to biogeochemical parameters.

Applications of bio-optical research include:

- **Ocean color remote sensing**: Interpreting satellite data (e.g., from MODIS, VIIRS, Sentinel-3 OLCI) to monitor chlorophyll, sediment plumes, and harmful algal blooms.
- **Primary productivity estimation**: Quantifying photosynthetic rates and carbon fluxes.
- **Ecosystem monitoring**: Detecting changes due to climate variability and anthropogenic impacts.
- **Marine resource management**: Informing policies on fisheries, aquaculture, and coastal development.
- **Validation of oceanographic models**: Supporting physical-biological coupled models in predicting coastal dynamics.

6 Indian Contributions and the Need for Integration

- India has made significant progress in oceanographic research through organizations such as the National Institute of Oceanography (NIO), the Indian National Centre for Ocean Information Services (INCOIS), the Space Applications Centre of ISRO (SAC-ISRO), and various academic institutions. However, much of the bio-optical data remains fragmented. This book seeks to address that gap by compiling and synthesizing region-specific research findings.
- Initiatives such as the Ocean Colour Monitor (OCM) aboard the Oceansat series and recent field campaigns under the Coastal Ocean Monitoring and Prediction System (COMAPS) have established a strong foundation for bio-optical calibration and validation in Indian waters. However, there remains a pressing need for high-resolution, long-term datasets, especially in under-studied regions such as the eastern coast, the Andaman Sea, and the Lakshadweep Islands.

7 Scope of This Book

This volume seeks to provide a comprehensive overview of the bio-optical properties of Indian coastal waters. It brings together contributions from leading researchers and practitioners, addressing topics such as:

- Spectral absorption and scattering of coastal waters.
- Variability of chlorophyll and phytoplankton groups.
- Optical effects of suspended and dissolved matter.
- Development and validation of regional bio-optical algorithms.
- Case studies from different coastal zones (e.g., Cochin coastal water-Arabian Sea, Hugli estuary-northern Bay of Bengal).
- Remote sensing applications in water quality monitoring and algal bloom detection.

Each review chapter/chapter provides insights into different coastal zones of Indian coastal water that collectively enhance our understanding of the complex interplay between light and matter in India's coastal ocean. The following are the snapshots of each chapter:

Chapter 2 highlights the importance of satellite-retrieved chlorophyll-a in monitoring Indian coastal waters. It evaluates algorithms, sensor performance, and applications in ecosystem health, algal blooms, and fisheries. Emphasizing regional calibration, it underlines the growing role of remote sensing in marine ecological monitoring and understanding biogeochemical variability.

Chapter 3 depicted that India's coastal waters show complex bio-optical variability due to freshwater influx, productivity, and physical processes. The Arabian Sea and Bay of Bengal differ in light absorption by substances like phytoplankton and CDOM. This study explores these dynamics, validates region-specific algorithms, and supports applications like algal bloom detection and marine pollution monitoring.

Chapter 4 presents a comprehensive study of phytoplankton dynamics and optical properties in the South Eastern Arabian Sea. Using multi-seasonal data (2010–2017), it explores environmental influences, CDOM variability, and bio-optical modeling, emphasizing the role of remote sensing in monitoring coastal ecosystems and improving marine resource management.

Chapter 5 explains that the north-western Bay of Bengal's bio-optical properties are shaped by riverine inputs, with seasonal shifts driven by sediment and CDOM during monsoon and phytoplankton dominance in non-monsoon periods. Increasing anthropogenic pressures highlight the need for improved regional models to enhance remote sensing accuracy in these complex coastal waters.

Chapter 6 highlights the significance of phytoplankton size structure and functional types in Indian coastal waters. It emphasizes their ecological roles, environmental influences, and impacts of human activities. Understanding these dynamics is crucial for evaluating marine ecosystem health, productivity, and informing effective conservation and management strategies in coastal environments.

In Chapter 7, the Hugli estuary, influenced by tides and seasonal monsoons, exhibits dynamic hydrographic and ecological conditions. Variations in salinity, temperature, and nutrient flux affect productivity and fish aggregation. Satellite data reveal seasonal changes in TSM, CDOM, and Chl-a levels, aiding in identifying Potential Fishing Zones crucial for sustainable fishery management.

Chapter 8, this review explores the dynamics of Chromophoric Dissolved Organic Matter (CDOM) along India's 7,500 km coastline. It highlights sources, spatial–temporal variability, and ecological roles across diverse ecosystems. Emphasizing CDOM's value as a proxy for water quality and carbon cycling, the study calls for standardized monitoring and region-specific research to guide sustainable management.

Chapter 9 describes that India has made significant progress in Ocean Colour Remote Sensing (OCRS) through satellite missions, research infrastructure, and capacity-building programs. Training initiatives and academic integration have empowered over 2500 participants, promoting OCRS use in oceanographic studies. Continued efforts are needed to develop region-specific algorithms and strengthen international collaborations.

8 Conclusion

As environmental pressures on coastal regions continue to escalate—driven by climate change, sea level rise, and intensifying anthropogenic activities—the demand for accurate, timely, and region-specific oceanographic data has become more urgent than ever. Coastal zones are among the most productive and vulnerable ecosystems on the planet, and understanding their changing dynamics is essential for effective environmental management and policy-making.

In this context, bio-optical studies play a crucial role. Situated at the intersection of physics, biology, and remote sensing, bio-optics provides vital insights into the interactions between light and water constituents, enabling the monitoring of key ecological processes such as primary productivity, water quality, and biogeochemical fluxes. These studies are indispensable for interpreting satellite-derived ocean color data and for improving the accuracy of models that inform decision-making in coastal resource management.

This book seeks to advance the field of marine optics by synthesizing existing knowledge, highlighting region-specific challenges, and identifying critical gaps in our understanding of Indian coastal waters. By doing so, it aims to inspire further research tailored to the unique environmental conditions of the Indian subcontinent.

I hope that the insights and analyses presented in this volume will contribute meaningfully to the sustainable management of India's coastal and marine ecosystems. In particular, I envision this work supporting efforts in marine spatial planning, habitat conservation, fisheries management, and climate adaptation—ultimately helping to build a more resilient and informed approach to coastal stewardship in the face of ongoing environmental change.

References

Carder, K. L., & Steward, R. G. (1985). A remote-sensing reflectance model of a red-tide dinoflagellate off West Florida. *Limnology and Oceanography, 30*(2), 286–298.

Carder, K. L., Steward, R. G., Harvey, G. R., & Ortner, P. B. (1989). Marine humic and fulvic acids: Their effects on remote sensing of ocean chlorophyll. *Limnology and Oceanography, 34*(1), 68–81.

Darecki, M., & Stramski, D. (2004). An evaluation of MODIS and SeaWiFS bio-optical algorithms in the Baltic Sea. *Remote Sensing of Environment, 89*(3), 326–350.

Das, S., Das, I., Giri, S., Chanda, A., Maity, S., Lotliker, A. A., Kumar, T. S., Akhand, A., & Hazra, S. (2017). Chromophoric dissolved organic matter (CDOM) variability over the continental shelf of the northern Bay of Bengal. *Oceanologia, 59*(3), 271–282.

Gordon, H. R., & Morel, A. Y. (1983). *Remote assessment of ocean color for interpretation of satellite visible imagery: A review. Lecture notes on coastal and estuarine studies* (Vol. 4). Springer-Verlag.

Hedges, J. I., & Keil, R. G. (1995). Sedimentary organic matter preservation: An assessment and speculative synthesis. *Marine Chemistry, 49*(2–3), 81–115. https://doi.org/10.1016/0304-4203(95)00008-F

Hochman, H., Vodacek, A., & Blough, N. V. (1994). The distribution of chromophoric dissolved organic matter (CDOM) in the Orinoco River plume. *Proceedings of the International Geoscience and Remote Sensing Symposium (IGARSS), 1994*(3), 1690–1692.

Kahru, M., & Mitchell, B. G. (1998). Spectral reflectance and absorption of a massive red tide off Southern California. *Journal of Geophysical Research: Oceans, 103*(C10), 21601–21609.

Morel, A., & Prieur, L. (1977). Analysis of variations in ocean color. *Limnology and Oceanography, 22*(4), 709–722.

Muller-Karger, F. E., Walsh, J. J., Evans, R. H., & Meyers, M. B. (1991). On the seasonal phytoplankton concentration and sea surface temperature cycles of the Gulf of Mexico as determined by satellites. *Journal of Geophysical Research: Oceans, 96*(C7), 12645–12665. https://doi.org/10.1029/91JC00787

Platt, T., Gallegos, C. L., & Harrison, W. G. (1988). Ocean primary production and available light: Further algorithms for remote sensing. *Deep-Sea Research Part A. Oceanographic Research Papers, 35*(6), 855–879. https://doi.org/10.1016/0198-0149(88)90064-7

Sathyendranath, S. (Ed.). (2000). *Remote sensing of ocean colour in coastal and other optically-complex waters. IOCCG Report No. 3* (p. 140). International Ocean-Colour Coordinating Group.

Stumpf, R. P., Frayer, M. L., Durako, M. J., & Brock, J. C. (1999). Variations in water clarity and bottom albedo in Florida bay from 1985 to 1997. *Estuaries, 22*(2B), 431–444.

Tassan, S. (1994). Local algorithms using SeaWiFS data for the retrieval of phytoplankton, pigments, suspended sediment, and yellow substances in coastal waters. *Applied Optics, 33*(12), 2369–2378.

Wollast, R. (1991). The coastal organic carbon cycle: Fluxes, sources, and sinks. In R. F. C. Mantoura, J.-M. Martin, & R. Wollast (Eds.), *Ocean margin processes in global change* (pp. 365–381). Wiley.

Chapter 2
A Comprehensive Review of Ocean Colour Satellite-Retrieved Chlorophyll and Associated Bio-Optical Algorithms in Indian Coastal Waters

Suchismita Srichandan, Amit Kumar Jena, Susmita Raulo, Sanjiba Kumar Baliarsingh, Alakes Samanta, Sudheer Joseph, T. M. Balakrishnan Nair, Kamal Kumar Barik, and Sourav Das

Abstract Marine ecological monitoring is essential for understanding and managing ocean ecosystems. The advent of ocean colour satellites has made remote sensing a powerful tool for assessing marine ecosystems on a large spatio-temporal scale. This chapter comprehensively reviews ocean colour satellite-retrieved chlorophyll-*a* (chl-*a*) and related bio-optical algorithms in Indian coastal waters. It covers the use of satellite-retrieved chl-*a* for research, retrieval algorithms, performance assessment, and applications, addressing their challenges and limitations. The review highlights the major use of satellite-retrieved chl-*a* in monitoring marine ecosystem health, tracking algal blooms, and understanding biogeochemical processes of Indian coastal waters, with significant operational applications in identifying potential fishing zones. A scientometric analysis (1990–2024) of satellite-retrieved chl-*a* in Indian coastal waters shows the growing interest and importance of using satellite data. Among the 154 publications screened, approximately 65% focused on the western coast of India and 35% on the eastern coast. Satellite-retrieved chl-*a* analysis has successfully tracked long-term phytoplankton trends, the evolution, decline of blooms, the spatial spread of algal blooms, impacts of extreme events on

S. Srichandan
Department of Marine Sciences, Berhampur University, Bhanja Bihar, Odisha, India

A. K. Jena
Indian Institute of Technology Bombay, Mumbai, India

S. Raulo · S. K. Baliarsingh (✉) · A. Samanta · S. Joseph · T. M. B. Nair
Indian National Centre for Ocean Information Services, Ministry of Earth Sciences, Government of India, Hyderabad, India

K. K. Barik
Centurion University of Technology and Management, Bhubaneswar, India

S. Das
Department of Chemistry, Brainware University, Kolkata, West Bengal, India

© The Author(s), under exclusive license to Springer Nature Switzerland AG 2026
S. Das (ed.), *Bio-optical Properties of Indian Coastal Waters*,
https://doi.org/10.1007/978-3-032-08714-0_2

algal productivity, and algal size distribution. Regarding sensor and algorithm performance, studies indicate that MODIS-Aqua and its sensor-default algorithm OC3M perform well across the entire Indian coastal waters. The recent OLCI ocean colour sensor onboard Sentinel-3A, using a neural network-based algorithm, has shown promising results for chl-*a* retrieval, as evidenced by research in the eastern Arabian Sea's coastal waters. Evaluations of satellite and biophysical model-retrieved chl-*a* data demonstrate strong correlations with field measurements, emphasizing the effectiveness of combining satellite and model data for understanding biogeochemical variability and primary productivity. This review emphasizes the need for regional bio-optical algorithms validated with regional calibration and validation datasets, considering seasonal and regional variations and their integration with existing monitoring programmes.

Keywords Ocean colour satellites · Remote sensing · Bio-optical properties · Chlorophyll-*a* · Phytoplankton bloom

1 Introduction

Coastal waters are among the planet's most biologically diverse regions, encompassing estuaries, mangroves, coral reefs, and seagrass beds, which support a vast array of species. This biodiversity is crucial for several reasons (Hinrichsen, 2013). These areas are highly productive, often called the "nurseries of the sea" due to their support for young marine life. Their productivity is essential for multiple reasons such as (1) coastal waters are key sites of primary production, driven by phytoplankton, seagrasses, and mangroves that form the base of the marine food web, (2) coastal regions are vital for global fisheries, providing a significant portion of the world's seafood, thereby supporting local economies and food security, (3) coastal vegetation, such as mangroves and seagrasses, helps sequester carbon dioxide, mitigating climate change (Short & Short, 1984).

Intensive monitoring of coastal waters is essential to maintain marine ecosystem health, protect public health, support economic activities, and prepare for climate change and natural disasters (Hernández-Delgado, 2015). Continuous and comprehensive monitoring efforts are necessary to address the complex challenges coastal regions face and ensure their sustainable future. Such monitoring is critical for environmental protection, public health, economic stability, biodiversity conservation, pollution control, habitat preservation, and managing harmful algal blooms.

Technological advancements, such as remote sensing, have improved the efficiency and accuracy of coastal water monitoring (Lotliker et al., 2021). Ocean Colour Remote Sensing (OCRS) is a vital tool for studying marine ecosystems. By observing ocean colour from space, scientists can gather crucial information about various biological, chemical, and physical processes in the marine environment. Key aspects of OCRS include monitoring phytoplankton, assessing water quality, supporting fisheries management, and enhancing oceanographic models (Baliarsingh et al., 2022). Important ocean colour parameters (including derived) include chlorophyll-*a* (chl-*a*), Total Suspended Matter (TSM), diffused attenuation coefficient (kd_{490}), Particulate Organic Carbon (POC), Coloured Dissolved Organic Mater (CDOM), phytoplankton absorption coefficient (a_{ph}), and detritus absorption coefficient (a_{dg}). Among these, chl-*a* is essential for understanding phytoplankton biomass, primary productivity, nutrient dynamics, and marine ecosystem health (Kramer & Siegel, 2019) and the highest explored parameter from ocean colour satellites.

Chl-*a* concentration is widely considered a proxy for phytoplankton biomass, which is crucial for understanding marine ecosystems and the modulation of global climate. Phytoplankton, the microscopic photosynthetic organisms in the sunlit layers of oceans, play a critical role in maintaining the balance of marine ecosystems and global biogeochemical cycles. Despite their diminutive size, phytoplankton contribute approximately half of the world's primary production, making them integral to the planet's carbon cycle and overall climate regulation (Falkowski et al., 1998). Their importance extends beyond primary production; phytoplankton form the foundation of aquatic food webs, supporting a diverse array of marine life, from zooplankton to large marine mammals.

The retrieval of chl-*a* concentration from OCRS satellite data is crucial for understanding marine ecosystems, particularly in coastal regions where anthropogenic and natural factors influence water quality (Brewin et al., 2010). Chl-*a* serves as a proxy for primary productivity and is an essential indicator of ocean health and nutrient dynamics (Hu et al., 2012). Accurate assessment of chl-*a* concentration is thus paramount for monitoring the biological and ecological state of marine environments.

Bio-optical algorithms are fundamental tools employed to derive chl-*a* concentrations from remote sensing data. These algorithms utilize the relationship between ocean colour, determined by water constituents' absorption and scattering properties, and chl-*a* concentration. The algorithms can be broadly classified into empirical, semi-analytical, and analytical types, each with distinct advantages and limitations (O'Reilly et al., 1998; Lee et al., 2002).

Empirical algorithms, such as the Ocean Chlorophyll 4 (OC4) algorithm, are based on statistical relationships between in situ chl-*a* measurements and satellite-derived reflectance ratios. These algorithms are straightforward and computationally efficient but may lack robustness in optically complex waters, such as those found in coastal regions (O'Reilly et al., 1998). Semi-analytical algorithms, on the other hand, incorporate both empirical relationships and inherent optical properties (IOPs) of water, offering improved performance in varying optical conditions (Lee et al., 2002).

The Indian coastal waters present a unique challenge for chlorophyll retrieval due to their complex optical properties influenced by high TSM, CDOM, and varying phytoplankton communities (Shanmugam, 2011). These waters require bio-optical algorithms that can accurately account for the diverse optical constituents and provide reliable chl-*a* estimates. Recent advancements in satellite sensor technology, such as the Ocean and Land Colour Instrument (OLCI) on Sentinel-3, have provided enhanced spectral resolution and sensitivity, facilitating the development of more sophisticated bio-optical algorithms (Bourg et al., 2015).

Against the above backdrop, this review aims to provide a comprehensive analysis of the bio-optical algorithms utilized for chl-*a* retrieval from ocean colour satellites over Indian coastal waters. By examining various algorithms' performance, advantages, and limitations, this chapter seeks to identify the most effective methods for monitoring chl-*a* concentrations in this region. Such an analysis is crucial for improving the accuracy of remote sensing applications in coastal monitoring and management.

2 Indian Coastal Water Vis-à-Vis OCRS of Chlorophyll-*a*

Indian coastal waters are ecologically significant, serving as habitats for diverse marine life and offering essential ecosystem services (Shenoi, 2024). These waters are crucial for fisheries, supporting millions of livelihoods and contributing significantly to the national economy (Nayak, 2017). They also play a key role in carbon sequestration, mitigating climate change by absorbing carbon dioxide. The aesthetic value of these areas promotes tourism, boosting the hospitality and recreation industries. However, these ecosystems face challenges from natural and human-induced forces. Issues like pollution, algal blooms, eutrophication, overfishing, habitat destruction, and climate change degrade water quality and marine health (Nair et al., 2024). Effective monitoring and sustainable management are essential to maintain the ecological integrity of Indian coastal waters (Sivadas et al., 2021).

OCRS has been crucial in monitoring the health of marine ecosystems, tracking algal blooms, and understanding biogeochemical processes in India (Mishra et al., 2022). It aids fisheries management by identifying productive fishing zones and supports coral reef conservation. This technology also provides valuable data for managing coastal pollution and addressing climate change impacts (Sahu et al., 2012; Baliarsingh et al., 2022).

Numerous studies have explored the variability of ocean colour parameters on large and regional scales (Chauhan & Raman, 2017). As the use of OCRS grows among the Indian oceanographic community, reviewing the effectiveness of different bio-optical algorithms is essential. These studies enhance our understanding of the complex dynamics of Indian coastal waters, underscoring the need for continued research and technological advancements.

In the Indian context, chl-*a* data from ocean colour satellites is extensively used for generating Potential Fishing Zone (PFZ) advisories and monitoring algal blooms (Sahu et al., 2012; Baliarsingh et al., 2022; Samanta et al., 2023). Chl-*a*, along with sea surface temperature (SST), serves as an indicator of pelagic fish availability and abundance, helping locate major fishing grounds, thereby reducing search time, fuel consumption, and expenses while increasing fishermen's catch (Solanki et al., 2003).

3 Evolution of Ocean Colour Satellite Sensors from Past to Present with Its Technological Advances

The journey of OCRS began in the 1970s with the launch of early satellite-sensors like the Coastal Zone Colour Scanner (CZCS) aboard Nimbus-7 in 1978 (Fig. 2.1). CZCS was a pioneering instrument that provided the first comprehensive view of the ocean's biological productivity (McClain, 2009). Despite its limitations in spatial resolution and calibration issues, CZCS demonstrated the potential of satellite-based OCRS observations.

Following CZCS, the Sea-viewing Wide Field-of-view Sensor (SeaWiFS) launch in 1997 marked a significant milestone. SeaWiFS offered improved spatial and spectral resolution, longer mission duration, and better radiometric calibration, leading to more reliable and extensive datasets (Hooker et al., 1992). SeaWiFS

Fig. 2.1 Timeline 1976–2024 illustrating past and current ocean colour satellite missions. The vertical axis lists the sensor/platform names and the horizontal axis indicates their respective operational periods

played a pivotal role in establishing long-term ocean colour records and enhancing our understanding of oceanic ecological processes.

Early sensors like CZCS had limited spectral bands and relatively coarse spatial resolution. Modern sensors, such as the Moderate Resolution Imaging Spectroradiometer (MODIS) on Terra and Aqua satellites and the Visible Infrared Imaging Radiometer Suite (VIIRS) on Suomi NPP, offer multiple spectral bands ranging from the visible to the near-infrared region. This increased spectral resolution allows for better discrimination of different water constituents, including chl-*a*, CDOM, and TSM (Werdell et al., 2013).

Radiometric calibration ensures that the sensor's measurements are accurate and stable over time. Advances in onboard calibration techniques, including the use of solar diffusers and lunar observations, have significantly improved the radiometric accuracy and stability of modern sensors. This is critical for maintaining the long-term consistency of OCRS datasets (Franz et al., 2007).

The development of sophisticated algorithms for atmospheric correction, which removes the effects of aerosols and water vapour, has been crucial for accurate OCRS data retrievals. Modern algorithms can account for varying atmospheric conditions and improve the accuracy of derived OCRS products. Additionally, machine learning and artificial intelligence have opened new avenues for processing and analysing large volumes of satellite data (IOCCG, 2014).

The frequency of observations has increased with the launch of multiple satellite missions. Sensors like MODIS and VIIRS provide near-daily global coverage, enabling continuous monitoring of dynamic oceanic phenomena. The Sentinel-3 mission of the European Space Agency offers enhanced temporal resolution with a revisit time of less than 2 days (Donlon et al., 2012).

Combining data from multiple sensors has become a common practice to enhance the spatial, spectral, and temporal resolution of OCRS products. For instance, the synergy between MODIS, VIIRS, and Sentinel-3 sensors allows for more comprehensive monitoring and cross-validation of data, leading to higher confidence in the observations (Gregg & Casey, 2004).

Today, OCRS is an emerging field with robust satellite programmes in place. The recent Plankton, Aerosol, Cloud, ocean Ecosystem (PACE) mission by NASA represents the next generation of OCRS sensors. PACE features the Ocean Colour Instrument (OCI), which promises unprecedented spectral resolution with capabilities extending into the ultraviolet and shortwave infrared regions (Werdell et al., 2019). This facilitates advanced studies on phytoplankton diversity, ocean carbon cycling, and climate interactions. The recent Indian OCRS satellite Oceansat-3, which houses an Ocean Colour Monitor (OCM) sensor aboard, has been observed to provide promising data records.

The evolution of OCRS satellite sensors from the early days of CZCS to today's advanced systems highlights the remarkable progress in remote sensing technology. These advancements have significantly improved our ability to monitor and understand the world's oceans, providing valuable insights into marine ecosystems and their response to environmental changes. As technology continues to advance, future OCRS missions will likely offer even greater capabilities, contributing to the sustainable management and conservation of marine resources.

4 General Global Observation So Far on the Performance of OCRS Algorithms in Retrieving Chlorophyll-*a* from Coastal Waters

The retrieval of chl-*a* concentrations from coastal waters using OCRS algorithms has been the subject of extensive research. Coastal waters present unique challenges for remote sensing due to their complex optical properties influenced by various factors, including a higher load of TSM, multiple sources of CDOM, and phytoplankton diversity.

OCRS algorithms developed for open ocean conditions often perform poorly in coastal waters due to the latter's optically complex nature. Algorithms like OC3 and OC4, which are designed for oligotrophic waters, tend to overestimate chl-*a* concentrations in turbid coastal waters (Wang et al., 2020). This discrepancy has led to the development of region-specific algorithms such as the OC2 and OC3 for the Chesapeake Bay (Harding et al., 2005).

The performance of these algorithms can vary significantly by region and season. For instance, Hu et al. (2000) demonstrated that algorithms calibrated for the Baltic Sea exhibit better performance in that region compared to globally calibrated algorithms. Seasonal changes, such as phytoplankton blooms, further complicate the accurate retrieval of chl-*a* concentrations.

Coastal waters often contain high concentrations of non-algal particles and CDOM, which can interfere with chl-*a* retrieval. The presence of these substances can lead to significant errors in algorithms not specifically designed to account for them (Gitelson et al., 2007). Multi-spectral and hyperspectral sensors have shown promising results in addressing these challenges by providing additional spectral information. Recent advancements have focused on machine learning approaches and bio-optical models that can better handle the complexity of coastal waters. For example, neural network algorithms have been developed to improve chl-*a* estimation by learning from large datasets of in situ measurements and satellite observations (Gitelson et al., 2007). A global analysis highlighted that while global algorithms provide a useful baseline, regional adaptations are crucial for accurate chl-*a* retrieval in coastal waters, which emphasized the need for continuous validation and calibration across different coastal regions (Wang et al., 2020).

5 Background of Genesis and Use of OCRS Algorithms

OCRS is based on the optical properties (apparent and inherent) of the ocean, and thus spectral values of the water leaving radiance are used in the bio-optical algorithms for retrieval of ocean colour parameters, including chl-*a* (Loisel et al., 2013). All the algorithms are developed considering the absorption, scattering, and reflectance of the light by different optically active substances present in ocean water in their respective wavelengths. In the context of chl-*a*, this pigment represents prominent features of strong absorption in blue (450–475 nm) and red (~670 nm) and peak reflectance at green (~550 nm) and near-infrared (~700 nm) of the visible spectrum (Gholizadeh et al., 2016). Additionally, chl-*a* emits a small part of the absorbed light at longer wavelengths as fluorescence, acting as an additional signal for remote sensing (Morel, 1980). Sometimes, the estimation of chl-*a* is hindered by the presence of CDOM and detritus. CDOM and detritus absorb a high amount of light with a decrease in wavelength in between 350 and 700 nm (Devi et al., 2015). CDOM absorption can overlap with the chl-*a* absorption as CDOM absorption is at its peak at blue (443 nm), which is the wavelength at which usually the chl-*a* is measured (Corbett, 2007). The chl-*a* is estimated based on the empirical relationships between reflectance in narrow bands (band ratio: blue to green) and chl-*a* concentration (Devi et al., 2015; Harrington & Repic, 1995). Therefore, the semi-analytical algorithms advantageously use a_{Ph}, CDOM, and detritus while retrieving chl-*a*.

6 Evolution of the OCRS Bio-Optical Algorithms for Retrieval of Chlorophyll-*a*

Several OCRS algorithms have been developed to retrieve chl-*a* concentrations. Most of these algorithms come under the category of empirical and semi-analytical types. During the initial phase, chl-*a* was retrieved using colour index algorithm, which was based on three bands (green, blue, and red) reflectance differences (Hu et al., 2019). Later, OCx algorithm was used, which is based on the fourth order polynomial relationship between reflectances and chl-*a*. Most of the available chl-*a* retrieval algorithms are based on the empirical OCx algorithm, where "x" is the number of the bands required for the retrieval of chl-*a*. The OCx algorithms use the maximum band ratio of remote sensing reflectances in blue and green regions of the spectrum. The bands used for the OCx algorithm vary depending on the ocean colour sensors and the availability of bands in which the sensors measure the reflectances. Attempts have been made to validate the chl-*a* estimation by various ocean colour sensors (e.g. MERIS, SeaWiFS, MODIS, OCM, VIIRS, and Sentinel) with the in situ measurements to understand the performance of the respective bio-optical algorithms and further tuning (Harshada et al., 2021). The details about various sensor-default bio-optical algorithms are provided in Table 2.1.

Table 2.1 Functional forms of various sensor-default bio-optical algorithms used for retrieving chlorophyll-*a* from different ocean colour satellites

Sensor	Algorithm	Functional form	Model type
CZCS	OC3C MBR	$$R = \log_{10}\left\{\max\left[\left(\frac{R_{rs}443}{R_{rs}555}\right),\left(\frac{R_{rs}520}{R_{rs}555}\right)\right]\right\}$$ $a = [0.31841; -4.56386; 8.63979; -8.41411; 1.91532]$ $$C = 10^{\left(a_0 + a_1R + a_2R^2 + a_3R^3 + a_4R^4\right)}$$	Fourth Order Polynomial
OLCI	OC4 MBR	$$R = \log_{10}\left\{\max\left[\left(\frac{R_{rs}443}{R_{rs}560}\right),\left(\frac{R_{rs}490}{R_{rs}560}\right),\left(\frac{R_{rs}510}{R_{rs}560}\right)\right]\right\}$$ $a = [0.42540; -3.21679; 2.86907; -0.62628; -1.09333]$ $$C = 10^{\left(a_0 + a_1R + a_2R^2 + a_3R^3 + a_4R^4\right)}$$	Fourth Order Polynomial
OCTS	OC4O MBR	$$R = \log_{10}\left\{\max\left[\left(\frac{R_{rs}443}{R_{rs}565}\right),\left(\frac{R_{rs}489}{R_{rs}565}\right),\left(\frac{R_{rs}516}{R_{rs}565}\right)\right]\right\}$$ $a = [0.54655; -3.51799; 3.39128; -0.91567; -0.97112]$ $$C = 10^{\left(a_0 + a_1R + a_2R^2 + a_3R^3 + a_4R^4\right)}$$	Fourth Order Polynomial
SeaWiFS	OC4 MBR	$$R = \log_{10}\left\{\max\left[\left(\frac{R_{rs}443}{R_{rs}555}\right),\left(\frac{R_{rs}489}{R_{rs}555}\right),\left(\frac{R_{rs}510}{R_{rs}555}\right)\right]\right\}$$ $a = [0.32814; -3.20725; 3.22969; -1.36769; -0.81739]$ $$C = 10^{\left(a_0 + a_1R + a_2R^2 + a_3R^3 + a_4R^4\right)}$$	Fourth Order Polynomial
MERIS	OC4E MBR	$$R = \log_{10}\left\{\max\left[\left(\frac{R_{rs}443}{R_{rs}560}\right),\left(\frac{R_{rs}489}{R_{rs}560}\right),\left(\frac{R_{rs}510}{R_{rs}560}\right)\right]\right\}$$ $a = [0.42487; -3.20974; 2.89721; -0.75258; -0.98259]$ $$C = 10^{\left(a_0 + a_1R + a_2R^2 + a_3R^3 + a_4R^4\right)}$$	Fourth Order Polynomial
MODIS	OC3M MBR	$$R = \log_{10}\left\{\max\left[\left(\frac{R_{rs}443}{R_{rs}547}\right),\left(\frac{R_{rs}488}{R_{rs}547}\right)\right]\right\}$$ $a = [0.26294; -2.64669; 1.28364; 1.08209; -1.76828]$ $$C = 10^{\left(a_0 + a_1R + a_2R^2 + a_3R^3 + a_4R^4\right)}$$	Fourth Order Polynomial
OCM2	OC4O2 MBR	$$R = \log_{10}\left\{\max\left[\left(\frac{R_{rs}443}{R_{rs}555}\right),\left(\frac{R_{rs}490}{R_{rs}555}\right),\left(\frac{R_{rs}510}{R_{rs}555}\right)\right]\right\}$$ $a = [0.475, -3.029, 2.240, -1.253, -0.027]$ $$C = 10^{\left(a_0 + a_1R + a_2R^2 + a_3R^3\right)} + a_4$$	Third Order Modified Cubic Polynomial

(continued)

Table 2.1 (continued)

Sensor	Algorithm	Functional form	Model type
VIIRS-SNPP	OC3_ VIIRS_ SNPP, CI	$$R = \log_{10}\left\{\max\left[\left(\frac{R_{rs}443}{R_{rs}551}\right),\left(\frac{R_{rs}486}{R_{rs}551}\right)\right]\right\}$$ $a = [0.23548; -2.63001; 1.65498; 0.16117; -1.37247]$ $$C = 10^{\left(a_0+a_1R+a_2R^2+a_3R^3+a_4R^4\right)}$$	Fourth Order Polynomial
VIIRS-NOAA20	OC3_ VIIRS_ NOAA20, CI	$$R = \log_{10}\left\{\max\left[\left(\frac{R_{rs}445}{R_{rs}556}\right),\left(\frac{R_{rs}489}{R_{rs}556}\right)\right]\right\}$$ $a = [0.28153; -2.65472; 1.30882; 1.31521; -2.08622]$ $$C = 10^{\left(a_0+a_1R+a_2R^2+a_3R^3+a_4R^4\right)}$$	Fourth Order Polynomial
VIIRS-NOAA21	OC3_ VIIRS_ NOAA21, CI	$$R = \log_{10}\left\{\max\left[\left(\frac{R_{rs}445}{R_{rs}555}\right),\left(\frac{R_{rs}488}{R_{rs}555}\right)\right]\right\}$$ $a = [0.24765; -2.54926; 1.55323; 0.39485; -1.54632]$ $$C = 10^{\left(a_0+a_1R+a_2R^2+a_3R^3+a_4R^4\right)}$$	Fourth Order Polynomial
OCEANSAT-3	EOS-06 (OCM)	$\mathrm{Log}_{10}(C) = a + b * R + c * R^2 + d * R^3$ $$R = \log_{10}\left\{\max\left[\left(\frac{R_{rs}443}{R_{rs}555}\right),\left(\frac{R_{rs}490}{R_{rs}555}\right),\left(\frac{R_{rs}510}{R_{rs}555}\right)\right]\right\}$$ $a = 0.2604$, $b = -2.8025$, $c = 3.6626$, and $d = -1.976$	Cubic Polynomial
PACE OCI	OC4 MBR	$$R = \log_{10}\left\{\max\left[\left(\frac{R_{rs}442}{R_{rs}555}\right),\left(\frac{R_{rs}490}{R_{rs}555}\right),\left(\frac{R_{rs}510}{R_{rs}555}\right)\right]\right\}$$ $a = [0.32814; -3.20725; 3.22969; -1.36769; -0.81739]$ $$C = 10^{\left(a_0+a_1R+a_2R^2+a_3R^3+a_4R^4\right)}$$	Fourth Order Polynomial
GOCI	OC4 MBR	$$R = \log_{10}\left\{\max\left[\left(\frac{R_{rs}412}{R_{rs}555}\right),\left(\frac{R_{rs}443}{R_{rs}555}\right),\left(\frac{R_{rs}489}{R_{rs}555}\right)\right]\right\}$$ $a = [0.28043; -2.49033; 1.53980; -0.09926; -0.68403]$ $$C = 10^{\left(a_0+a_1R+a_2R^2+a_3R^3+a_4R^4\right)}$$	Fourth Order Polynomial
HAWKEYE	OC4 MBR	$$R = \log_{10}\left\{\max\left[\left(\frac{R_{rs}447}{R_{rs}556}\right),\left(\frac{R_{rs}488}{R_{rs}556}\right),\left(\frac{R_{rs}510}{R_{rs}556}\right)\right]\right\}$$ $a = [0.32814, -3.20725; 3.22969; -1.36769; -0.81739]$ $$C = 10^{\left(a_0+a_1R+a_2R^2+a_3R^3+a_4R^4\right)}$$	Fourth Order Polynomial

Source: www.earthdata.nasa.gov

7 Synthesis on the Satellite-Retrieved Chlorophyll-*a* and Associated Retrieval Algorithms in Indian Coastal Waters

Indian coastal water comprises coastal waters of the western Bay of Bengal (east coast) and eastern Arabian Sea (west coast). Accounting for both the east and west coasts of India, the ocean colour satellite-retrieved chl-*a* in Indian coastal waters has become a vital tool for monitoring and understanding marine ecosystems, including its use in the generation of several related operational ecological services. Starting in the ~2000s, studies using ocean colour satellite-retrieved chl-*a* have focused on various aspects, including regular monitoring of chl-*a* concentration, tracking algal blooms, assessing the impact of tropical cyclones in fuelling phytoplankton upsurge, analysing regional as well as basin-scale long-term trends, deciphering the response of phytoplankton communities to environmental changes, estimating primary productivity, utilization as input in ocean models, and utilization in locating potential fishing grounds (Fig. 2.2). The recent decade has witnessed significant advancements in the use of ocean colour satellite-retrieved chl-*a* in studying phytoplankton types, characterizing algal blooms, analysing spatio-temporal extent of different phytoplankton size classes, etc. Research in this domain has highlighted the importance of using accurate retrieval algorithms and validation with in situ measurements to enhance data reliability. Despite challenges related to algorithm accuracy and data validation, the advancements in satellite technology continue to improve the precision and applicability of chl-*a* measurements, making them indispensable for marine research and environmental monitoring in the Indian coastal waters.

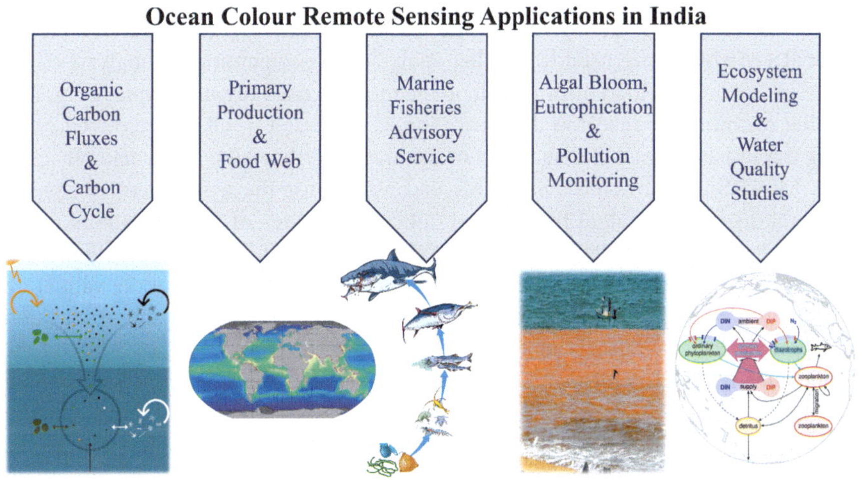

Fig. 2.2 Image showing the major application of ocean colour satellite-retrieved chlorophyll-*a* in India

8 Scientometric Analysis of Scientific Research w.r.t Satellite-Retrieved Chlorophyll-*a* in the Indian Coastal Waters

The scientometric analysis of scientific research concerning satellite-retrieved chl-*a* in Indian coastal waters provides a comprehensive overview of this field's trends, patterns, and impacts. By examining the volume of publications and collaborative networks, this analysis highlights the growing interest and significance of using satellite technology to monitor chl-*a*. The findings reveal key research hotspots and prolific authors contributing to this domain, offering insights into the evolution of methodologies and the integration of OCRS in oceanography. Such an analysis not only underscores the advancements in satellite technology and its applications but also identifies research gaps.

A systematic review was conducted using information from 1990 to 2024 (till July). The process involved three steps: (1) selecting articles via search engines (using Elsevier's Scopus, Google Scholar, Academia, ProQuest, PubMed, etc.); (2) screening articles based on relevant research topics; and (3) extracting relevant information. This yielded 354 articles. The search targeted scientific articles, reviews, and book chapters on OCRS of chl-*a* in the Indian coastal waters. Keywords included "Chlorophyll-*a*", "Algorithm", "Ocean colour", "Remote sensing", "coastal", "Satellite", "Arabian Sea", "Bay of Bengal", and "Indian ocean" focusing on India. Duplicate records were removed after additional bibliography checks. Titles and abstracts were examined, and ultimately, 154 publications were reviewed.

From the screened 154 publications, nearly ~65% of studies were pertaining to the western coast of India and ~35% from the east coast. The scientometric analysis was carried out using VOSviewer software, focusing on the author and indexed keywords of all 154 publications. The keywords having a repeatability of more than five were considered, out of which the keywords representing the research domains of OCRS of chl-*a* were used for further analysis. The scientometric analysis of the filtered keywords yielded information on the most focused research topics (Fig. 2.3) and the evolution of research domains (Fig. 2.4) related to OCRS retrieved chl-*a* along the Indian coastal waters. Most of the studies utilized the chl-*a* data retrieved from MODIS-Aqua, SeaWiFS, OCM, and VIIRS for the assessment of spatio-temporal distribution, algal bloom, and optical properties of Case-II waters in the earlier years of introduction of OCRS to the oceanographic society. With the development of the technologies, the research domain gradually shifted to the validation of chl-*a* retrieval algorithms, development of regional algorithms, and association of chl-*a* with other optically active substances. The recent studies focusing on the application of satellite-retrieved chl-*a* for the identification of potential fishing zones, detection of algal blooms, identification of phytoplankton groups (diatom and dinoflagellate) and size classes, and estimation of primary productivity in the Indian coastal waters.

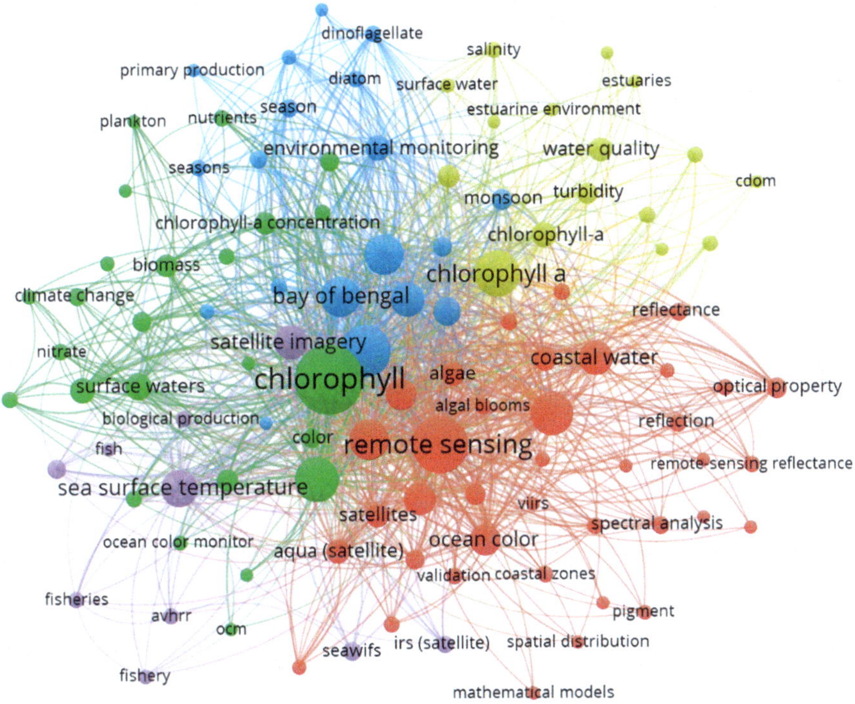

Fig. 2.3 Scientometric network visualizing the research topics using satellite-retrieved chlorophyll-*a* in the Indian coastal waters

8.1 Case Study of the Bay of Bengal: Monitoring

8.1.1 Investigating the Influence COVID-19 Lockdown on Coastal Water Quality

In order to assess the impact of the COVID-19 lockdown on Indian coastal water quality parameters in Indian coastal waters, Lotliker et al. (2021) used MODIS-Aqua to analyse water quality parameters, including chl-*a*, POC, and kd_{490}. As the North Indian Ocean exhibits seasonality in the spatio-temporal variation of water quality parameters, an anomaly-based assessment was adopted in this study to eliminate the seasonal biases. Focusing on ten strategic locations along the Indian coastline, the study found a significant reduction in these parameters during the lockdown, especially in the western Bay of Bengal, attributed to decreased anthropogenic nutrient input. The coastal waters off major cities like Chennai and river basins like Hooghly exhibited substantial negative anomalies, indicating improved water clarity. This reduction is linked to the decrease in river discharge carrying industrial, agricultural, and urban effluents. The paper highlights the effectiveness of satellite remote sensing in environmental monitoring and emphasizes the importance of

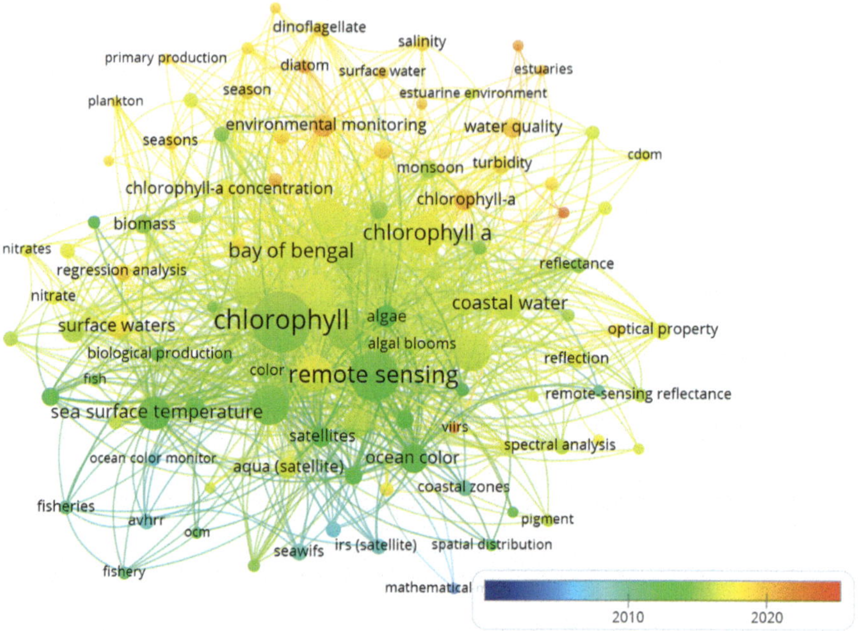

Fig. 2.4 Scientometric network plot showing year-wise evolution of research domains of marine ecology using satellite-retrieved chlorophyll-*a* in the Indian coastal waters

continuous monitoring to understand the long-term impacts of anthropogenic activities on coastal ecosystems.

8.1.2 Estimation of Phytoplankton Size Classes

The research by Miranda et al. (2021) on phytoplankton size classes (PSC) in the coastal waters of north-western Bay of Bengal is notable for its use of OCRS data to estimate and validate phytoplankton size distributions over time. The study validated two models, Brewin et al. (2010) (B10) and Sahay et al. (2017) (S17), finding the S17 model more accurate. Spatial distribution analysis revealed nearshore dominance of microphytoplankton and offshore dominance of picophytoplankton, with seasonal variations showing microphytoplankton prevailing except during the southwest monsoon. The research highlighted higher microphytoplankton concentrations during pre-monsoon due to recurrent blooms and post-monsoon due to nutrient influx from precipitation (Fig. 2.5).

Methodologically, the study used in situ data to validate MODIS-Aqua satellite data, demonstrating the S17 model's superior performance with lower relative errors. The study recommends the S17 model for satellite PSC retrieval in this region, emphasizing the value of satellite data in understanding phytoplankton dynamics and broader ocean climate interactions. This work sets a foundation for

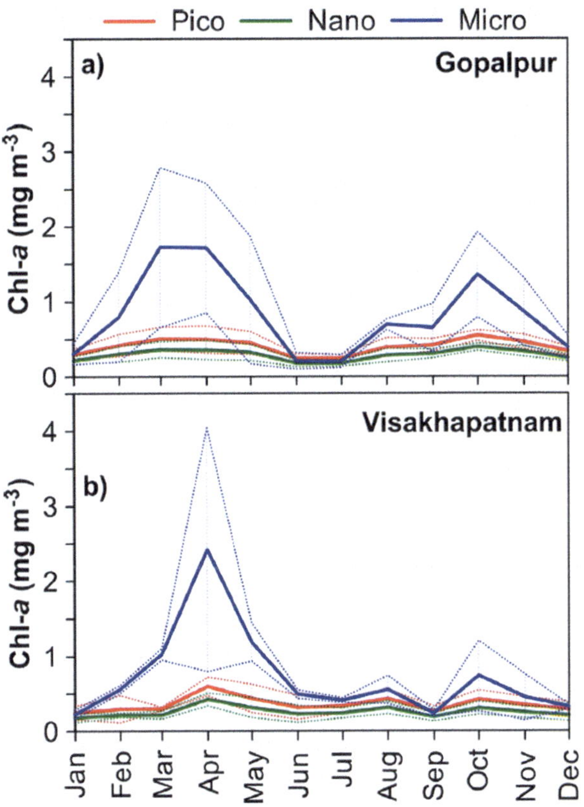

Fig. 2.5 Area averaged monthly climatological variability (2002–2018) of satellite-retrieved (using S17 model) chlorophyll-*a* (chl-*a*) concentration of pico, nano, and microphytoplankton in coastal waters of (**a**) Gopalpur and (**b**) Visakhapatnam. The solid lines represent the mean, and the dotted lines represent the standard deviation. (Source: Miranda et al., 2021)

future research on marine ecosystems and climate change impacts, providing critical insights into phytoplankton behaviour and its environmental drivers.

8.1.3 Characterization of Phytoplankton Blooms

A recent study by Srichandan et al. (2022) examined OCRS-based characterization of phytoplankton blooms in the coastal waters of the north-western bay. Diatom blooms, the most common, are frequent in this region. For the first time, the study investigated the long-term trends in OCRS satellite-retrieved diatom distribution. Retrospective analysis of bloom events matched earlier in situ observations, revealing a bi-modal annual trend with peaks during pre- and post-southwest monsoon periods. This pattern is attributed to nutrient availability and favourable environmental conditions. Diatom distribution showed a more extensive spread in post-monsoon. The study also linked diatom availability to the proliferation of heterotrophic red *Noctiluca scintillans*. Intense blooms of diatoms, dinoflagellates, and cyanobacteria have been reported, indicating susceptibility to coastal water quality disruption. Diatom blooms have been observed for over two decades, and this study is the first to evaluate long-term trends using advanced satellite remote sensing. The study highlighted the efficacy of OCRS in deciphering the long-term trends of chl-*a* and diatom distribution and retrospective analysis of algal bloom events. The concordance of in situ observed diatom blooms with elevated chl-*a* concentration matched with satellite data. It is important to mention that this study provided a clear evidence of diatoms pre-conditioning water for *Noctiluca scintillans* proliferation. This study advocates for developing species-specific satellite retrieval algorithms and advanced ocean colour sensors to capture short-term changes in chl-*a*, crucial for understanding the impact of blooms on ecosystem functioning, especially under global climate change scenarios.

8.2 Case Study of the Bay of Bengal: Performance Assessment

8.2.1 Inter-Sensor Comparison and Long-Term Dynamics

The study by Lotliker et al. (2020) investigated the long-term dynamics of chl-*a* in coastal waters of the north-western bay. This research focuses on the distribution of optically active substances such as chl-*a*, TSM, and CDOM, using in situ data and satellite retrievals from MODISA, VIIRS, and OCM-2 sensors. The key findings highlight significant spatial and temporal variability in chl-*a*, with prominent bimodal peaks occurring during the pre-southwest monsoon (March–April) and towards the end of the southwest monsoon (August–October). The pre-monsoon peak is attributed to recurrent phytoplankton blooms, while the monsoon peak is linked to nutrient influx from river runoff. The study found that MODISA provided the most accurate satellite retrievals of chl-*a* although all sensors showed some overestimation in nearshore waters (Fig. 2.6). The study's analogy inferred that the complex interactions between physical processes, such as river discharge, monsoon-driven circulation, and local upwelling, in shaping the distribution of chl-*a*. These findings highlight the importance of long-term satellite data in understanding the ecological variability and biogeochemical processes in coastal marine environments.

8.2.2 Semi-Analytical Algorithm and Associated IOP

Lotliker et al. (2015) conducted an investigation into the IOP of the north-western bay, examining their variability and implications for OCRS. The study evaluated two semi-analytical algorithms, Garver-Siegel-Maritorena (GSM) and Generalized IOP (GIOP), and their relationship with IOPs such as a_{ph}, a_{dg}, and particulate backscattering (b_{bp}). The research utilized 10 years of chl-*a* data retrieved from

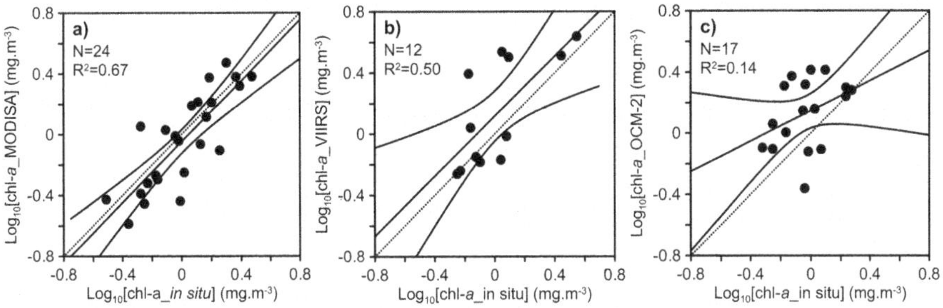

Fig. 2.6 Scatterplot showing the relation between in situ measured chlorophyll-*a* and that derived from (**a**) MODISA, (**b**) VIIRS, and (**c**) OCM-2. The dotted line indicates 1:1, the solid straight line shows the regression and the solid curved line shows the 95% confidence limit. (Source: Lotliker et al., 2020)

MODIS-Aqua using the GSM and GIOP algorithms. Both algorithms exhibited similar trends on an annual scale, but the magnitude of chl-*a* from GSM was three times higher than that from GIOP. Moreover, the chl-*a* retrieved from GSM was closer to in situ values (Fig. 2.7).

The long-term variability of a_{ph} and a_{dg} was similar, with a peak during pre-monsoon attributed to bloom occurrences and a peak during monsoon due to nutrient discharge from rivers. The weak association between IOP and chl-*a* highlights the limited accuracy of satellite-derived chl-*a* from these semi-analytical algorithms. This study recommended generating regional-scale IOPs and fine-tuning the semi-analytical model is necessary for improved accuracy.

8.2.3 Assessment of Chlorophyll-*a* Retrieval Algorithms for OCM-2

The chl-*a* data retrieved from OCM-2 using OC2, OC3, and OC4 algorithms were compared to see the retrieval accuracy of the algorithms in the coastal waters of the Bay of Bengal (Paradip, Mahanadi, and Haldia) (Nagamani et al., 2013). The OC4 and OC3 algorithms were found to overestimate the chl-*a*, whereas OC2 performed better in the coastal waters of Paradip, Mahanadi, and Hugli estuary (Haldia). In another study, the SeaWiFS chl-*a* data obtained using the OC5, OC4v6, and Carder (both empirical and semi-analytical) algorithms were evaluated for the coastal waters of the bay (Tilstone et al., 2011). The assessment, based on in situ chl-*a* measurements, revealed that the OC5 algorithm performed better in these waters compared to OC4v6 and Carder. While OC4v6 was more accurate than Carder, OC4v6 overestimated chl-*a* concentrations above 2 mg/m^3, whereas Carder overestimated chl-*a* above 1 mg/m^3 and underestimated it below 0.5 mg/m^3. When comparing the three algorithms, OC5 and OC4v6 showed a better correlation than Carder. The study recommended the OC5 algorithm for chl-*a* retrieval in the bay's coastal waters due to its non-covariance with other optically active substances like TSM and CDOM. In contrast, OC4v6 and Carder algorithms exhibited covariance with TSM, significantly impacting chl-*a* retrieval in coastal waters.

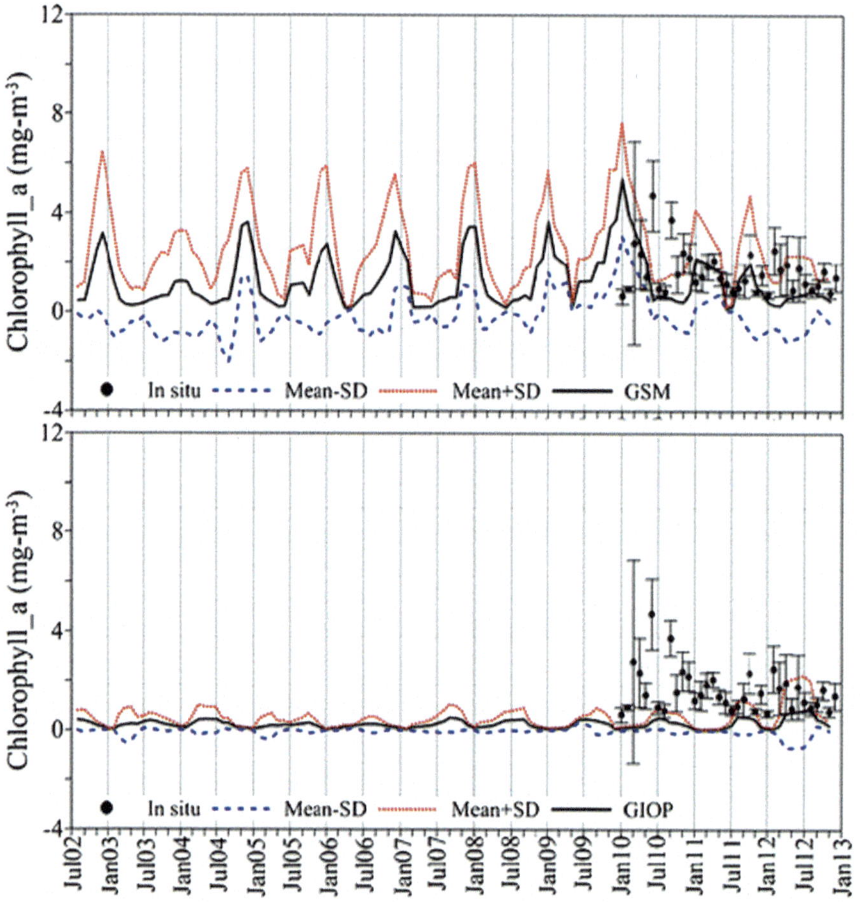

Fig. 2.7 Time-series variability in the monthly mean distribution of chl-*a* derived from MODIS-Aqua using GSM (upper chart) and GIOP (lower chart) algorithms. The dotted line (red) represents the sum of the Mean and standard deviation (SD). The dashed line (blue) represents the difference between Mean and SD. The dots represent the in situ chl-*a*, and the vertical lines represent the standard deviation. (Source: Lotliker et al., 2015)

8.3 Case Study of the Arabian Sea: Monitoring

8.3.1 Mapping Coastal Green *Noctiluca* Bloom

An investigation by Samanta et al. (2023) monitored a significant algal bloom event along the southwestern coast of India caused by green *Noctiluca scintillans* (Fig. 2.8). This intense bloom resulted in water discolouration, a scummy surface layer, bioluminescent tides, and mass fish mortality. This study used the Algal Bloom Information Service (ABIS) of the Indian National Centre for Ocean Information Services, which uses several OCRS products, including chl-*a*, and

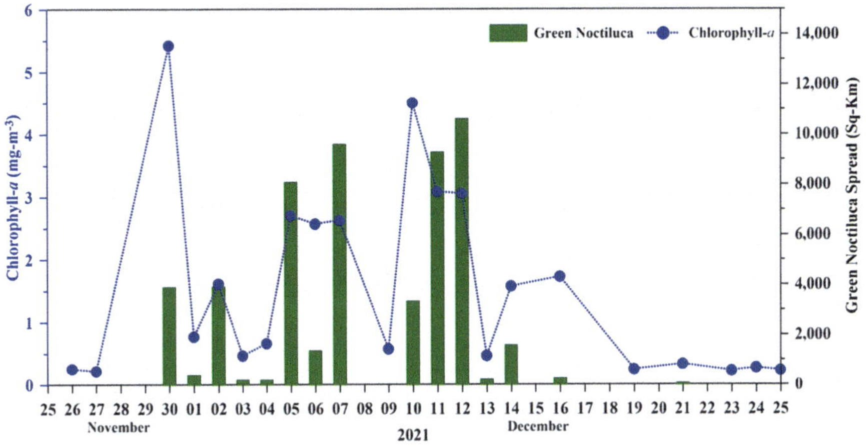

Fig. 2.8 Variability of green *Noctiluca scintillans* spread area and chlorophyll-*a* concentration in the study area. (Source: Samanta et al., 2023)

monitored the bloom from its onset on November 30, 2021, to its peak on December 12, 2021, and eventual decline on December 21, 2021. The chl-*a* concentrations of 5.42, 3.01, and 0.33 mg/m^3 were recorded during the bloom's initiation, peak, and decline, respectively, and *Noctiluca* spread areas of 3888, 10,608, and 48 sq.km during these periods. The study emphasizes the importance of OCRS in monitoring and understanding the factors causing such blooms and their ecological impacts.

8.4 Case Study of the Arabian Sea: Performance Assessment

8.4.1 Assessment of Chlorophyll-*a* Retrieval Algorithms

The algorithms OC3M, GSM, and GIOP, developed for the retrieval of chl-*a*, were assessed using MODIS-Aqua data and validated with in situ measurements (Tilstone et al., 2013). The study found that OC3M outperformed GSM and GIOP in the coastal waters of the Arabian Sea. OC3M's retrieval accuracy was within 11% compared to in situ chl-*a*, while GSM's accuracy was within 24% and GIOP's was 55%. Additionally, OC3M was less affected by errors in remote sensing reflectance and by spectral variations in CDOM and TSM compared to the other algorithms.

OC4E algorithm as compared to the in situ chl-*a* in the coastal waters of Kerala (eastern Arabian Sea), provided the underestimation of ~66%, 51%, and 44% by the sensors respectively (Nimit et al., 2016). From the study, it was observed that OC4E and OC3M offered better chl-*a* retrieval as compared to MERIS-CC. Another study by Minu et al. (2016) evaluated the empirical algorithms (OC3C, OC4O, OC4, OC4E, OC3M, and OC4O2) for the retrieval of chl-*a* from CZCS, OCTS, SeaWiFS, MERIS, MODIS, and OCM-2 in the coastal waters of Kerala. The evaluation showed that although the performance of all the algorithms was poor in the coastal waters but OC3M and OC4 had relatively better performance than other algorithms.

An inter-comparison of ocean colour sensors on the estimation of chl-*a* revealed an overestimation of 25–32% by MODIS-Aqua when compared to SeaWiFS in the coastal waters of Karnataka (eastern Arabian Sea). Meanwhile, the inter-comparison between MERIS and SeaWiFS was found to give similar estimates of chl-*a* (Arun Kumar et al., 2015). Thus, the validation of ocean colour sensors and algorithms is necessary for complex coastal waters.

8.4.2 Inter-Sensor Comparison

Harshada et al. (2021) compared chl-*a* data from various sensors—MODIS, SNPP-VIIRS, OC-CCI, and Sentinel-3A OLCI—with in situ measurements from the coastal waters of Karnataka in the southeastern Arabian Sea. They found that the Sentinel-3A OLCI FR (C2RCC OLCI processed) chl-*a* product at 300 m, which uses a neural network-based algorithm, closely matched the field data, making it the most accurate for the coastal waters in the study area. The study noted an uncertainty of 33%, indicating that the neural network approach effectively accounts for other optical constituents in the coastal waters. For products with coarser resolutions, uncertainty increased to 39.9%, highlighting the significance of spatial resolution. Data with a 4 km resolution performed poorly compared to higher resolution data, primarily due to spatial binning of adjacent pixels and the use of blue-green reflectance ratio algorithms. MODIS data at 4 km resolution was less effective in coastal waters than at 1 km resolution, while SNPP-VIIRS chl-*a* data, derived using a ratio algorithm at 750 m, showed better accuracy compared to the 1 km resolution data.

8.4.3 Evaluation of Satellite and Model-Retrieved Chlorophyll-*a*

The study by Chakraborty et al. (2016) evaluated the accuracy of sea surface chl-*a* concentrations retrieved from the MODIS-Aqua satellite using the OC3M algorithm and from a Regional Ocean Modelling System (ROMS) biophysical model against in situ measurements in the eastern Arabian Sea from April 2009 to December 2012. This study indicated that MODIS-Aqua chl-*a* concentrations show a high correlation with in situ data, with a slope close to unity and low root mean square error, although there was a marginal underestimation by the satellite data (Fig. 2.9).

In contrast, the ROMS model underestimated chl-*a* but still captured the overall trends. The study also examined time-series trends in chl-*a* against wind-driven upwelling indices, finding that the ROMS model accurately captured the annual peak in chl-*a* during the summer monsoon along the Kochi coast, driven by coastal upwelling and the winter monsoon peak along the Gujarat coast due to convective mixing. Despite some underestimations, the ROMS model effectively reproduced

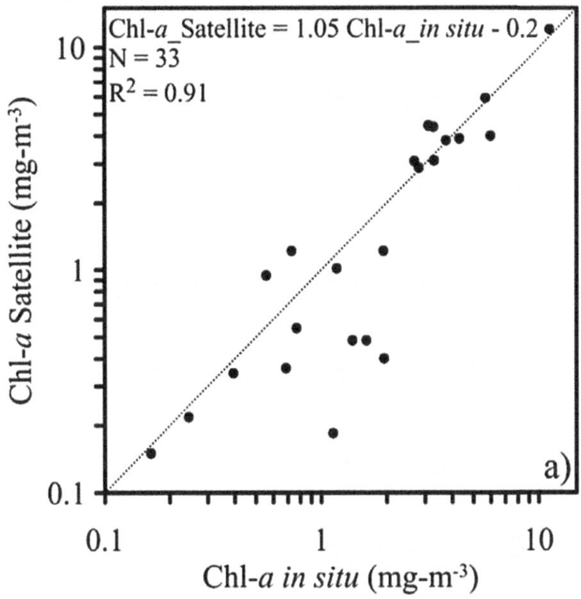

Fig. 2.9 Log-log plot of MODIS-Aqua OC3M chl-*a* versus in situ chl-*a*. The scale represents the actual values. The dotted line represents 1:1. (Source: Chakraborty et al., 2016)

the seasonal and inter-annual trends in chl-*a*, demonstrating its capability to simulate the dynamics of phytoplankton blooms in response to monsoonal upwelling and mixing processes in the Arabian Sea. The study emphasizes the utility of combining satellite and model data to improve our understanding of biogeochemical variability and primary productivity in coastal marine environments.

8.5 Comparative Account of the Performance of Key Ocean Colour Satellite Sensors in Retrieving Chlorophyll-a

The available literature review suggests that a majority of the studies using ocean colour data are based on regular data products based on sensor-default algorithms. In general, at regular intervals, comparative studies between in situ and satellite-derived chl-*a* measurements are necessary and pivotal to assess the performance of the retrieval algorithms. Based on the assessment, the needful tuning is carried out to enhance the performance and development of new algorithms.

In this context, Table 2.2 depicts an overview of validation exercises carried out to assess the performance of sensor-default algorithms of critical ocean colour satellite sensors in the coastal waters of India and other parts of the global oceans. For Indian waters, the comparative studies presented the validation result of MODIS-Aqua with in situ chl-*a* through R^2 ranging from 0.67 to 0.91. For instance, a study from the Bay of Bengal shows an R^2 of 0.67 along with a Log10 RMSE of 0.18, with errors in the MODIS-Aqua satellite data ranging from 33% to 51%. Another study from the Northern Bay of Bengal, with R^2 of 0.78, suggests the moderate reliability of the MODIS-Aqua may be due to the bio-optical complexities in the coastal waters of the Bay of Bengal. In contrast, MODIS-Aqua performed better in the Arabian Sea with an R^2 of 0.91 and a Log10 RMSE of 0.24, reflecting good alignment with the in situ observation. Thus, MODIS-Aqua was able to be a useful tool in providing insights into the long-term variability of critical ecosystem health, including studies conducted during the COVID-19 pandemic across various Indian locations.

Other sensors, such as VIIRS, show mixed results, with R^2 values of 0.50 in the north-western Bay of Bengal and 0.523 in the southwest Bay of Bengal, with RMSE values of 0.26 and 0.846, respectively. The poor performance of VIIRS in the Indian coastal waters might be attributed to the inferred studies that have taken into account a relatively low number of data points for validation in comparison to the MODIS-Aqua. Considering the increasing use of VIIRS in recent times, future validation exercises may yield better results in proving the efficacy of VIIRS. The OCM sensor mainly used for Indian waters, also demonstrated a mixed result as in the north-western Bay of Bengal, OCM recorded a very low R^2 of 0.14, possibly due to limited in situ data points. Other studies also showed good performance of OCM in the Bay of Bengal, with R^2 of 0.821, and RMSE of 0.436. However, OCM data in the Arabian Sea have shown a strong correlation with in situ data with an R^2 of 0.90 and RMSE of 0.125.

Comparatively, studies conducted in other global coastal waters often achieved good correlations with the in situ observations, reflecting the better performance of the sensors in the respective regions, which may be due to less optical complexity than in Indian waters. For example, MODIS-Aqua recorded with R^2 of 0.83 and RMSE of 0.33 in the North Pacific. Yet, in the Chesapeake Bay, MODIS-Aqua wasn't able to retrieve the chl-*a,* represented by a sharp decline in R^2 (0.009) and high RMSE (23.217). In contrast, VIIRS achieved high correlation with in situ chl-*a* values in Chesapeake Bay, with R^2 of 0.95 (RMSE of 0.137) at 375-m resolution, and R^2 of 0.90 (RMSE of 0.207) at 750-m resolution, showing the importance of high resolution sensor in this region. The OCM sensor, which is mostly used for the Indian waters, also used for the chl-*a* retrieval in the northwest part of the Gulf of Thailand, yielded a good correlation (R^2 of 0.86 and RMSE of 0.88) with in situ data. This analysis reveals that the chl-*a* retrieval capability of different ocean colour satellites varies significantly across regions, likely due to local bio-optical conditions, such as suspended and organic material load. The analogy of Table 2.2 underpins the necessity of regional bio-optical algorithms and high-frequency validation-based tuning to assess and enhance performance.

Table 2.2 A comparative account of the performance of key ocean colour satellite sensors in retrieving chlorophyll-*a* using sensor-default bio-optical algorithms in coastal waters of India vis-a-vis coastal water worldwide

Indian coastal waters				
Sensors	Year	Region	Remarks	Reference
MODISA	2010–2012	Coastal waters of the north-western Bay of Bengal	In situ vs. satellite: R^2 of 0.67, Log_{10}RMSE 0.18	Lotliker et al. (2020)
	2003–2020	Coastal waters off major Indian cities (Mumbai and Chennai) and river basins	Long-term trend analysis provided critical insights describing the changes in the coastal ecosystem health w.r.t COVID-19 pandemic	Lotliker et al. (2021)
	2009–2012	Coastal water of the eastern Arabian Sea	In situ vs. satellite: R^2 of 0.91, and Log10 RMSE of 0.24	Chakraborty et al. (2016)
	2002–2018	Northern Bay of Bengal	In situ vs. satellite: R^2 of 0.78	Rouf et al. (2020)
VIIRS	2010–2012	Coastal waters of the north-western Bay of Bengal	In situ vs. satellite: R^2 of 0.50 and Log_{10}RMSE of 0.26	Lotliker et al. (2020)
	2013–2017	Southwest Bay of Bengal coast	In situ vs. satellite: R^2 of 0.523 and RMSE of 0.846	Priyanka et al. (2023)
OCM	2010–2012	Coastal waters of the north-western Bay of Bengal	In situ vs. satellite: R^2 of 0.14 and Log_{10}RMSE of 0.18	Lotliker et al. (2020)
	1999	Arabian Sea	In situ vs. satellite (OC2 algorithm): R^2 of 0.90 and RMSE of 0.125	Chauhan et al. (2002)
	2010–2012	Southwestern Bay of Bengal	In situ vs. satellite: R^2 of 0.92	Natesan et al. (2015)
	2010–2011	Southwestern Bay of Bengal	OCM2 vs. MODIS: R^2 of 0.549	Poornima et al. (2013)
	2011	North-Western Bay of Bengal	In situ vs. satellite: R^2 of 0.821 and RMSE of 0.436	Nagamani et al. (2013)

Other coastal waters (outside India)				
Sensors	Year	Region	Remarks	Reference
MODISA	2002–2014	North Pacific	In situ vs. satellite: R^2 of 0.83 and RMSE of 0.33	Carswell et al. (2017)
	2012–2016	Chesapeake Bay	In situ vs. satellite: R^2 of 0.009 and RMSE of 23.217	Abbas et al. (2019)
OCM	2016	The northwest part of the Gulf of Thailand	In situ vs. satellite: R^2 of 0.86 and RMSE of 0.88	Viswanath et al. (2018)

(continued)

Table 2.2 (continued)

Other coastal waters (outside India)				
Sensors	Year	Region	Remarks	Reference
VIIRS	2013	Chesapeake Bay	In situ vs. satellite (375-m resolution): R^2 of 0.95 and RMSE of 0.137 In situ vs. satellite (750-m resolution): R^2 of 0.90 and RMSE of 0.207	Vandermeulen et al. (2015)
	2011–2018	Chesapeake Bay	In situ vs. satellite: R^2 of 0.56 and RMSE of 4.7	Yu et al. (2022)

MODISA moderate resolution imaging spectroradiometer-Aqua, *VIIRS* visible infrared imaging radiometer suite, *OCM* Ocean colour monitor

9 Summary and Conclusion

In summary, this comprehensive review collectively underscores the role of satellite remote sensing in understanding coastal ecosystem responses to anthropogenic and climatic influences. It advocates for refined monitoring strategies to capture short-term variations crucial for ecosystem management amidst global climate change scenarios. The utilization of ocean colour satellites for remote sensing has revolutionized the ability to assess marine environments on extensive spatio-temporal scales. This comprehensive review of ocean colour satellite-retrieved chl-*a* and associated bio-optical algorithms in Indian coastal waters highlights significant advancements and applications in this domain.

The review emphasizes the pivotal role of satellite-retrieved chl-*a* data in monitoring marine ecosystem health, detecting algal blooms, and elucidating biogeochemical processes in Indian coastal waters. This technology has proven especially valuable for operational applications such as identifying potential fishing zones, thereby aiding in resource management and economic planning. A scientometric analysis spanning from 1990 to 2024 reveals a growing scholarly interest in the application of satellite data, with a notable concentration of research on the western coast of India compared to the eastern coast.

Key findings from the review indicate that satellite-retrieved chl-*a* data effectively track long-term phytoplankton trends, monitor the evolution and spatial spread of algal blooms, assess the impacts of extreme events on algal productivity, and analyse algal size distribution. Sensor performance evaluations demonstrate that MODIS-Aqua and its default algorithm OC3M are well suited for the diverse conditions of Indian coastal waters. The review highlights that while significant advancements have been made in the development of ocean colour sensors and bio-optical algorithms for chl-*a* estimation, there remains a need for region-specific calibration to improve their accuracy in Indian coastal waters. Continuous validation and refinement of these algorithms are essential to account for the unique optical characteristics of the region. Future research should focus on integrating multi-spectral and hyperspectral data to enhance the precision of chlorophyll retrieval.

The chapter underscores the critical role of satellite-based chl-*a* monitoring in supporting sustainable management and conservation of marine resources in the Indian coastal ecosystem. Most available algorithms are developed and validated using global in situ datasets measured for open ocean waters, which may not accurately estimate concentrations of optically active substances, including chl-*a*, in optically complex coastal waters. Therefore, algorithms need validation using regional in situ data and fine-tuning to develop regional algorithms for precise estimates in the coastal waters of India. While global algorithms provide a foundation, region-specific adaptations are crucial for improving accuracy. Advancements in sensor technology and algorithm development, including machine learning approaches, offer promising avenues for future research. Ongoing validation and regional calibration remain key to enhancing the reliability of chl-*a* retrieval from coastal waters. Future research should focus on integrating multi-spectral and hyperspectral data to improve algorithm performance. Continued efforts in the calibration of region-specific algorithms and the collection of in situ data will be essential for advancing our understanding and capability in this area.

Acknowledgements The authors express gratitude to the authorities of their respective affiliated institutions for supporting this piece of research. They also acknowledge all the published research that has been cited and reviewed in this chapter. The secondary data used in this study is from different published literature, which are properly cited with full reference. This paper constitutes a component of one of the author's doctoral research works (Susmita Raulo). This is INCOIS contribution no. 571.

References

Abbas, M. M., Melesse, A. M., Scinto, L. J., & Rehage, J. S. (2019). Satellite estimation of chlorophyll-a using moderate resolution imaging spectroradiometer (MODIS) sensor in shallow coastal water bodies: Validation and improvement. *Water (Basel), 11*(8), 1621.

Arun Kumar, S. V. V., Babu, K. N., & Shukla, A. K. (2015). Comparative analysis of chlorophyll-a distribution from SEAWIFS, MODIS-AQUA, MODIS-TERRA and MERIS in the Arabian Sea. *Marine Geodesy, 38*(1), 40–57.

Baliarsingh, S. K., Samanta, A., Lotliker, A. A., Mohanty, P. C., Mahendra, R. S., & Nair, T. B. (2022). Satellite-based marine ecological services for the Indian Ocean Region. In *Social and economic impact of earth sciences* (pp. 229–251). Springer Nature Singapore.

Bourg, L., Bernard, G., & Antoine, D. (2015). Ocean and land colour instrument (OLCI) on board Sentinel-3: The instrument, its capabilities and first data. *Ocean Science, 11*(6), 909–923.

Brewin, R. J., Hardman-Mountford, N. J., Lavender, S. J., & Hirata, T. (2010). An intercomparison of bio-optical techniques for detecting phytoplankton functional types from space. *Remote Sensing of Environment, 114*(3), 490–505.

Carswell, T., Costa, M., Young, E., Komick, N., Gower, J., & Sweeting, R. (2017). Evaluation of MODIS-aqua atmospheric correction and chlorophyll products of Western north American coastal waters based on 13 years of data. *Remote Sensing, 9*(10), 1063.

Chakraborty, K., Gupta, A., Lotliker, A. A., & Tilstone, G. (2016). Evaluation of model simulated and MODIS-aqua retrieved sea surface chlorophyll in the eastern Arabian Sea. *Estuarine, Coastal and Shelf Science, 181*, 61–69.

Chauhan, P., & Raman, M. (2017). Satellite remote sensing for ocean biology: An Indian perspective. *Proceedings of the National Academy of Sciences, India Section A: Physical Sciences, 87*, 629–640.

Chauhan, P., Mohan, M., Sarngi, R. K., Kumari, B., Nayak, S., & Matondkar, S. G. P. (2002). Surface chlorophyll a estimation in the Arabian Sea using IRS-P4 ocean colour monitor (OCM) satellite data. *International Journal of Remote Sensing, 23*(8), 1663–1676.

Corbett, C. A. (2007). *Colored dissolved organic matter (CDOM) workshop summary*. University of South Florida.

Devi, G. K., Ganasri, B. P., & Dwarakish, G. S. (2015). Applications of remote sensing in satellite oceanography: A review. *Aquatic Procedia, 4*, 579–584.

Donlon, C. J., et al. (2012). The global monitoring for environment and security (GMES) Sentinel-3 mission. *Remote Sensing of Environment, 120*, 37–57.

Falkowski, P. G., Barber, R. T., & Smetacek, V. (1998). Biogeochemical controls and feedbacks on ocean primary production. *Science, 281*(5374), 200–206.

Franz, B. A., Bailey, S. W., Werdell, P. J., & McClain, C. R. (2007). Sensor-independent approach to the vicarious calibration of satellite ocean color radiometry. *Applied Optics, 46*(22), 5068–5082.

Gholizadeh, M. H., Melesse, A. M., & Reddi, L. (2016). A comprehensive review on water quality parameters estimation using remote sensing techniques. *Sensors, 16*(8), 1298.

Gitelson, A. A., Gurlin, D., Moses, W. J., & Barrow, T. (2007). A bio-optical algorithm for the remote estimation of the chlorophyll-a concentration in coastal waters. *Remote Sensing of Environment, 109*(2), 142–155.

Gregg, W. W., & Casey, N. W. (2004). Global and regional evaluation of the SeaWiFS chlorophyll data set. *Remote Sensing of Environment, 93*(4), 463–479.

Harding, L. W., Magnuson, A., & Mallonee, M. E. (2005). SeaWiFS retrievals of chlorophyll in Chesapeake Bay and the mid-Atlantic bight. *Estuarine, Coastal and Shelf Science, 62*(1–2), 75–94.

Harrington, J., & Repic, R. (1995). Hyperspectral and video remote sensing of Oklahoma lakes. In *Papers and Proceedings of Applied Geography Conferences* (pp. 79–86). Applied Geography Conferences.

Harshada, D., Raman, M., & Jayappa, K. S. (2021). Evaluation of the operational chlorophyll-a product from global ocean colour sensors in the coastal waters, south-eastern Arabian Sea. *The Egyptian Journal of Remote Sensing and Space Science, 24*(3), 769–786.

Hernández-Delgado, E. A. (2015). The emerging threats of climate change on tropical coastal ecosystem services, public health, local economies and livelihood sustainability of small islands: Cumulative impacts and synergies. *Marine Pollution Bulletin, 101*(1), 5–28.

Hinrichsen, D. (2013). *Coastal waters of the world: Trends, threats, and strategies*. Island Press.

Hooker, S. B., Brown, J. W., & Balch, W. M. (1992). *SeaWiFS technical report series*. NASA.

Hu, C., Carder, K. L., & Muller-Karger, F. E. (2000). Atmospheric correction of SeaWiFS imagery over turbid coastal waters: A practical method. *Remote Sensing of Environment, 74*(2), 195–206.

Hu, C., Lee, Z., & Franz, B. (2012). Chlorophyll-a algorithms for oligotrophic oceans: A novel approach based on three-band reflectance difference. *Journal of Geophysical Research: Oceans, 117*(C1).

Hu, C., Feng, L., Lee, Z., Franz, B. A., Bailey, S. W., Werdell, P. J., & Proctor, C. W. (2019). Improving satellite global chlorophyll a data products through algorithm refinement and data recovery. *Journal of Geophysical Research: Oceans, 124*(3), 1524–1543.

IOCCG. (2014). Phytoplankton functional types from space. In *Reports of the International Ocean-colour coordinating group* (Vol. 15). IOCCG.

Kramer, S. J., & Siegel, D. A. (2019). How can phytoplankton pigments be best used to characterize surface ocean phytoplankton groups for ocean color remote sensing algorithms? *Journal of Geophysical Research: Oceans, 124*(11), 7557–7574.

Lee, Z. P., Carder, K. L., & Arnone, R. A. (2002). Deriving inherent optical properties from water color: A multiband quasi-analytical algorithm for optically deep waters. *Applied Optics, 41*(27), 5755–5772.

Loisel, H., Vantrepotte, V., Jamet, C., & Ngoc Dat, D. (2013). *Challenges and new advances in ocean color remote sensing of coastal waters*. INTECH.

Lotliker, A. A., Baliarsingh, S. K., Sahu, K. C., & Kumar, T. S. (2015). Performance of semianalytical algorithm and associated inherent optical properties in coastal waters of North Western Bay of Bengal. *Journal of the Indian Society of Remote Sensing, 43*, 143–149.

Lotliker, A. A., Baliarsingh, S. K., Sahu, K. C., & Kumar, T. S. (2020). Long-term chlorophyll-a dynamics in tropical coastal waters of the western Bay of Bengal. *Environmental Science and Pollution Research, 27*, 6411–6419.

Lotliker, A. A., Baliarsingh, S. K., Shesu, R. V., Samanta, A., Naik, R. C., & Balakrishnan Nair, T. M. (2021). Did the coronavirus disease 2019 lockdown phase influence coastal water quality parameters off major Indian cities and river basins? *Frontiers in Marine Science, 8*, 648166.

McClain, C. R. (2009). A decade of satellite ocean color observations. *Annual Review of Marine Science, 1*, 19–42.

Minu, P., Lotliker, A. A., Shaju, S. S., Ashraf, P. M., Kumar, T. S., & Meenakumari, B. (2016). Performance of operational satellite bio-optical algorithms in different water types in the southeastern Arabian Sea. *Oceanologia, 58*(4), 317–326.

Miranda, J., Lotliker, A. A., Baliarsingh, S. K., Jena, A. K., Samanta, A., Sahu, K. C., & Kumar, T. S. (2021). Satellite estimates of the long-term trend in phytoplankton size classes in the coastal waters of north-western Bay of Bengal. *Oceanologia, 63*(1), 40–50.

Mishra, P., Begum, M., Gera, A., Kumar, B. C., Deviram, G., Pradhan, U. K., Vashi, A., Bandyopadhyay, D., Naik, S., Panda, U. S., & Murthy, M. R. (2022). Factors responsible for the sudden outburst of Noctiluca scintillans in the Chennai coastal waters, southeast coast of India—a case study. *Oceanologia, 64*(4), 781–788.

Morel, A. (1980). In-water and remote measurements of ocean colour. *Boundary-Layer Meteorology, 18*, 177–181.

Nagamani, P. V., Hussain, M. I., Choudhury, S. B., Panda, C. R., Sanghamitra, P., Kar, R. N., Das, A., Ramana, I. V., & Rao, K. H. (2013). Validation of chlorophyll-a algorithms in the coastal waters of Bay of Bengal initial validation results from OCM-2. *Journal of the Indian Society of Remote Sensing, 41*, 117–125.

Nair, T. B., Sarma, V. V. S. S., Lotliker, A. A., Muraleedharan, K. R., Samanta, A., Baliarsingh, S. K., Shivaprasad, S., Gireeshkumar, T. R., Raulo, S., Vighneshwar, S. P., & Shesu, R. V. (2024). An integrated buoy-satellite based coastal water quality nowcasting system: India's pioneering efforts towards addressing UN Ocean decade challenges. *Journal of Environmental Management, 354*, 120477.

Natesan, U., Ferrer, V. A., Sahay, A., Chauhan, P., Muthulakshmi, A. L., Deepthi, K., Venugopalan, V. P., & Narasimhan, S. V. (2015). Monitoring chlorophyll–a concentration in the coastal waters of Kalpakkam (India): Application of a regional algorithm using Oceansat–2 ocean colour monitor. *Journal of the Indian Society of Remote Sensing, 43*, 173–179.

Nayak, S. (2017). Coastal zone management in India—present status and future needs. *Geo-Spatial Information Science, 20*(2), 174–183.

Nimit, K., Lotlikar, A., & Srinivasa Kumar, T. (2016). Validation of MERIS sensor's CoastColour algorithm for waters off the west coast of India. *International Journal of Remote Sensing, 37*(9), 2066–2076.

O'Reilly, J. E., et al. (1998). Ocean color chlorophyll algorithms for SeaWiFS. *Journal of Geophysical Research: Oceans, 103*(C11), 24937–24953.

Poornima, D., Shanthi, R., Raja, S., Sethubathi, G. V., Thangaradjou, T., Balasubramanian, T., Babu, K. N., & Shukla, A. K. (2013). Understanding the spatial variability of chlorophyll a and total suspended matter distribution along the southwest bay of Bengal using in-situ and OCM-2 & MODIS-aqua measurements. *Journal of the Indian Society of Remote Sensing, 41*, 651–662.

Priyanka, K., Sarangi, R. K., Shanthi, R., Louwinanand, D., & Saravanakumar, A. (2023). An empirical algorithm to estimate silicate in the Southwest Bay of Bengal. *Indian Journal of Geo-Marine Sciences (IJMS), 52*(09), 415–423.

Rouf, M. A., Antu, A. H., & Noor, I. (2020). Seasonal and annual variability in chlorophyll-in the shelf region of the Northern Bay of Bengal using MODIS-aqua data. *Oceanological and Hydrobiological Studies, 49*(4), 398–407.

Sahay, A., Ali, S. M., Gupta, A., & Goes, J. I. (2017). Ocean color satellite determinations of phytoplankton size class in the Arabian Sea during the winter monsoon. *Remote Sensing of Environment, 198*, 286–296.

Sahu, K. C., Baliarsingh, S. K., Srichandan, S., Lotliker, A. A., & Kumar, T. S. (2012). Validation of PFZ advisories—a case study along Ganjam coast of Odisha, East Coast of India. *Indian Streams Research Journal, 1*(XII), 1–6.

Samanta, A., Baliarsingh, S. K., Lotliker, A. A., Joseph, S., & Nair, T. B. (2023). Satellite-based detection of Noctiluca bloom in the coastal waters of the south-eastern Arabian Sea: A case study implicating monitoring needs. *National Academy Science Letters, 46*(2), 103–107.

Shanmugam, P. (2011). New models for retrieving and partitioning bio-optical properties in oceanic waters. *Journal of Geophysical Research: Oceans, 116*(C2).

Shenoi, S. S. C. (2024). Nurturing the blue economy: A call for sustainable ocean utilization. *Current Science, 126*(2).

Short, F. T., & Short, C. A. (1984). The seagrass filter: Purification of estuarine and coastal waters. In *The estuary as a filter* (pp. 395–413). Academic Press.

Sivadas, S. K., Muthukumar, C., Bharathi, M. D., Ramu, K., Srivastava, P. K., & Murthy, M. R. (2021). Connecting India's coastal monitoring program with UN sustainable development goal 14. *Ocean and Coastal Management, 215*, 105949.

Solanki, H. U., Dwivedi, R. M., Nayak, S. R., Somvanshi, V. S., Gulati, D. K., & Pattnayak, S. K. (2003). Fishery forecast using OCM chlorophyll concentration and AVHRR SST: Validation results off Gujarat coast, India. *International Journal of Remote Sensing, 24*(18), 3691–3699.

Srichandan, S., Baliarsingh, S. K., Samanta, A., Jena, A. K., Lotliker, A. A., Nair, T. B., Barik, K. K., & Acharyya, T. (2022). Satellite-based characterization of phytoplankton blooms in coastal waters of the Northwestern Bay of Bengal. *Journal of the Indian Society of Remote Sensing, 50*(11), 2221–2228.

Tilstone, G. H., Angel-Benavides, I. M., Pradhan, Y., Shutler, J. D., Groom, S., & Sathyendranath, S. (2011). An assessment of chlorophyll-a algorithms available for SeaWiFS in coastal and open areas of the Bay of Bengal and Arabian Sea. *Remote Sensing of Environment, 115*(9), 2277–2291.

Tilstone, G. H., Lotliker, A. A., Miller, P. I., Ashraf, P. M., Kumar, T. S., Suresh, T., Ragavan, B. R., & Menon, H. B. (2013). Assessment of MODIS-aqua chlorophyll-a algorithms in coastal and shelf waters of the eastern Arabian Sea. *Continental Shelf Research, 65*, 14–26.

Vandermeulen, R. A., Arnone, R., Ladner, S., & Martinolich, P. (2015). Enhanced satellite remote sensing of coastal waters using spatially improved bio-optical products from SNPP–VIIRS. *Remote Sensing of Environment, 165*, 53–63.

Viswanath, S. K., Tripathi, N. K., & Salin, K. R. (2018). Mapping of marine chl-a and suspended solid concentration using OCM-2 sensor. *Journal of the Indian Society of Remote Sensing, 46*, 675–685.

Wang, M., Son, S., & Harding, L. W. (2020). Satellite-observed algal blooms in the Chesapeake Bay and adjacent Atlantic Ocean. *Scientific Reports, 10*(1), 1–13.

Werdell, P. J., Bailey, S. W., Fargion, G. S., & McClain, C. R. (2013). *The SeaWiFS bio-optical archive and storage system (SeaBASS): Current architecture and implementation*. NASA Technical Memorandum.

Werdell, P. J., et al. (2019). The plankton, aerosol, cloud, ocean ecosystem mission: Status, science, advances. *Bulletin of the American Meteorological Society, 100*(9), 1775–1794.

Yu, X., Shen, J., Zheng, G., & Du, J. (2022). Chlorophyll-a in Chesapeake Bay based on VIIRS satellite data: Spatiotemporal variability and prediction with machine learning. *Ocean Modelling, 180*, 102119.

Chapter 3
Bio-Optical Depictions in the Coastal Waters of India: A Brief Review

Sudarsana Rao Pandi, Nittala S. Sarma, and Sarat C. Tripathy

Abstract The coastal waters of India undergo dynamic changes in their light absorption characteristics, influenced by a complex interplay of factors. In the northern Indian Ocean, characterized by two arms, namely the Arabian Sea (AS) and the Bay of Bengal (BOB), distinct bio-optical differences occur. The AS, experiencing less river influence and higher productivity, contrasts with the BOB, where substantial freshwater influx from river discharges significantly impacts light attenuation. This influx reduces light absorption by optically significant substances (OSS or OAS) like phytoplankton, detritus, and colored dissolved organic matter (CDOM), particularly in deeper waters. Additionally, the BOB coastal waters are further shaped by eddies and recurrent cyclonic storms, adding variability to OSS light attenuation properties. The study delves into the spatiotemporal variability of bio-optical constituents, aiming to understand the physical factors influencing light absorption by phytoplankton and related components across the coastal Indian Ocean. Notably, the investigation validates and develops algorithms specific to this region, highlighting differences in bio-optical properties between the AS and BOB. For instance, the specific absorption due to phytoplankton enables clear differentiation of phytoplankton size classes (PSCs) in coastal regions. Furthermore, CDOM absorption properties exhibit molecular weight, aromaticity, and conservative and non-conservative behaviors. These serve as tracers for identifying various sources of dissolved organic matter (DOM) in the coastal zone. During cyclonic and non-cyclonic events, the apparent optical properties of OSS exhibit significant variability, further emphasizing the complexity of coastal dynamics. Understanding

S. R. Pandi (✉)
National Centre for Polar and Ocean Research (NCPOR), Ministry of Earth Sciences, Vasco-da-Gama, Goa, India

Marine Chemistry Laboratory, Andhra University, Visakhapatnam, India

N. S. Sarma
Marine Chemistry Laboratory, Andhra University, Visakhapatnam, India

S. C. Tripathy
National Centre for Polar and Ocean Research (NCPOR), Ministry of Earth Sciences, Vasco-da-Gama, Goa, India

© The Author(s), under exclusive license to Springer Nature Switzerland AG 2026

S. Das (ed.), *Bio-optical Properties of Indian Coastal Waters*,
https://doi.org/10.1007/978-3-032-08714-0_3

OSS light absorption properties not only aids in estimating carbon levels but also contributes to comprehending the optical coherence resulting from physical processes in the AS and the diverse coastal BOB. This research sheds light on the intricate relationship between freshwater stratification, environmental dynamics, and bio-optical constituents, elucidating the optical properties of coastal waters in the region. This review also presents the recent advances made in the regionally tuned remote sensing algorithms for real-time monitoring, helping with objectives like biological resource mapping, algal bloom, including harmful algal bloom warnings and marine pollution episodes.

Keywords Hyperspectral remote sensing · Remote sensing reflectance · Particulate backscattering · Bay of Bengal · Arabian Sea

1 Introduction

Bio-optics is an essential modern tool for understanding the biological state and biogeochemistry of carbon in the ocean. Bio-optical properties involve the optically active substances or optically significant substances, i.e., substances expressed by the interaction with light in the ocean, such as phytoplankton, colored dissolved organic matter (CDOM), detritus, the medium of water itself, and other suspended particles (Oubelkheir et al., 2005).

Light plays a vital role in aquatic ecosystems, particularly as the primary energy source for phytoplankton photosynthesis. Through this process, phytoplankton absorb light and sequester atmospheric carbon dioxide, producing organic matter that supports marine food webs. Photosynthesis primarily occurs within the 400–700 nm range, known as photosynthetically active radiation (PAR), which is biologically useful and non-damaging. Shorter wavelengths are typically harmful to cells but are filtered by the ozone layer, while longer wavelengths lack sufficient energy to drive photosynthesis. However, some groups, such as cyanobacteria, can utilize light beyond this conventional range. PAR constitutes approximately 50% of incoming solar energy (Lozano et al., 2023).

Light penetration in the ocean varies with OAS in the upper water column. The $K_d(490 \text{ nm})$ is used to understand this penetration, as this wavelength typically experiences the least absorption in ocean waters. The euphotic depth (Z_{eu}), the depth at which only 1% of surface light remains provides a measure of the PAR. The first optical depth (τ), calculated as $Z_{eu}/4.6$ (Morel, 1988), represents the depth from which ~90% of the water-leaving radiance (detected by ocean color satellites) originates and is often approximated as $1/K_d$ (penetration depth, where ~37% of surface light remains). Additionally, Secchi depth, the depth at which a black-and-white Secchi disk disappears from view, is a simple and effective measure of water transparency or turbidity.

The light interaction by OAS occurs through various mechanisms, including absorption, transmission, reflection, and scattering of radiation (IOCCG, 2006; Tiwari & Kolluru, 2023). So the total solar irradiance is the sum of all these phenomena ($I_0 = I_{reflected} + I_{scattered} + I_{absorbed} + I_{transmitted}$), which is helpful to convert one property to another property via inverse mechanisms (Morel & Prieur, 1977; Morel & Maritorena, 2001). These light interaction properties can be classified into two types: inherent and apparent optical properties (Shifrin, 1998; D'Sa & Miller, 2005; Kirk, 2011; Mobley, 2022).

Inherent optical properties (IOP) refer to the inherent characteristics of a medium that affect the propagation of light within it. These properties are inherent to the medium itself and are independent of the specific path or conditions through which light travels (Gordon et al., 1975, Gordon & Morel, 1983; IOCCG, 2000; Mobley et al., 2004; Mobley, 2022). Examples of IOPs are the absorption coefficient, scattering coefficient, and phase function, which describe how light is absorbed and scattered within the medium without considering any external factors.

Apparent optical properties (AOP) are the characteristics of a medium as observed from the outside, accounting for the effects of both IOPs and external factors such as the path length of light, the distribution of particles within the medium, and the angle of observation (Mobley et al., 2004; Mobley, 2022). Unlike IOPs, AOPs are influenced by the specific conditions of measurement and are subject to variation based on factors such as depth, turbidity, and water composition. Examples of AOPs are the attenuation coefficient, beam attenuation, and R_{rs}, which provide information about how light interacts with the medium as observed from an external perspective.

The oceanic waters are classified into Case 1 and Case 2 based on their optical properties and biological characteristics (IOCCG, 2000; Mobley, 2022). Case 1 waters are characterized by clear water with relatively low concentrations of OAS, such as phytoplankton, suspended particles, and CDOM. In Case 1 waters typically occur in open ocean regions, away from coastal influences, where phytoplankton blooms are the primary source of OAS. Case 2 waters are characterized by higher concentrations of OAS, such as phytoplankton, suspended sediments, and CDOM. These waters are often found in coastal regions, estuaries, and areas influenced by river discharge, where input from terrestrial sources and nutrient-rich waters can lead to increased biological productivity and higher concentrations of OAS.

The OAS plays an important role in remote sensing of ocean color (Sosik & Mitchell, 1995; Menon et al., 2005; Stomp et al., 2007). The total absorption due to the Ocean (a_T) is a sum of absorptions by all OAS ($a_T = a_w + a_{CDOM} + a_D + a_{ph}$). Therefore, it is essential to apply corrections for all co-absorbing components, particularly a_{CDOM}, it can contribute 50% or more of the total absorption at 443 nm ($a(443)$) (Clementson et al., 2001; Bonelli et al., 2021), when estimating chlorophyll-*a* (Chl-*a*) and PP in the ocean.

1.1 Phytoplankton Light Absorption Properties

Phytoplankton sequester over 40% of anthropogenic CO_2 through primary production (PP), contributing significantly to climate regulation (Falkowski, 1994). Approximately 10% of this carbon is exported to the ocean interior as particulate and dissolved organic matter via the biological carbon pump (Siegel et al., 2014; Deppeler & Davidson, 2017). Light absorption during PP alters phytoplankton cell size and intracellular pigment composition (Ciotti et al., 2002; Lohrenz et al., 2003). At high Chl-*a* concentrations, the pigment packaging effect reduces absorption efficiency. Normalizing phytoplankton absorption spectra (a_{ph}) by Chl-*a* reveals size-related optical characteristics (Ciotti et al., 2002; Ferreira et al., 2017, Tripathy et al., 2020), showing a power-law relationship with Chl-a, modulated by the relative abundance of phytoplankton size classes (PSCs) (Babin et al., 2003; Bricaud et al., 2004). The a_{ph} also varies with light conditions: high irradiance can induce photoinhibition (Han et al., 2000), while low-light environments trigger photoacclimation (Ciotti et al., 2002; Fujiki & Taguchi, 2002; Tomkins et al., 2020).

Phytoplankton absorption is widely used to distinguish PSCs—micro, nano, and picophytoplankton (Sathyendranath et al., 2001; Devred et al., 2006; Hirata et al., 2008). Hirata et al. (2008) classified PSCs using absorption at 443 nm ($a_{ph}(443)$): $>0.06\ m^{-1}$ for micro, $0.01–0.06\ m^{-1}$ for nano, and $<0.01\ m^{-1}$ for picophytoplankton. Aguilar-Maldonado et al. (2017) further used the blue-to-red absorption ratio (B(443)/R(675)) to distinguish PSCs: >3.0 for pico, <2.5 for micro, and intermediate values for nanophytoplankton.

Beyond their optical signatures, phytoplankton also play a central role in coastal carbon cycling. During their growth phase, they utilize nutrients and dissolved organic matter (DOM) (Coble et al., 1998; Chari et al., 2013; Sarma et al., 2019), while their senescence and decay contribute to POM and detritus, which are further remineralized by bacterial and photochemical processes. These transformations influence DOM–POM fluxes and provide insights into ecosystem functioning and carbon remineralization. Understanding phytoplankton bio-optical properties is therefore essential not only for classifying community structure but also for assessing their role in biogeochemical cycles and photo-acclimation processes in coastal environments.

1.2 Phytoplankton Pigment Composition, Size Class, and Functional Types

The PSCs are classified, pico (size 0.2–2μm), nano (2–20μm), and micro (>20μm) phytoplankton based on their size (Sieburth et al., 1978), which can also be distinguished by their diagnostic marker pigments measured by High Performance Liquid Chromatography (HPLC). Models based on pigment composition, phytoplankton absorption, and R_{rs} have been developed to classify PSCs on synoptic scales. For instance, Uitz et al. (2006) utilized HPLC to quantify marker pigments. Several models have further contributed to PSCs estimation, including three-component models by Brewin et al. (2010, 2015), Pandi et al., (2025) and Sun et al. (2023), two-component models by Sathyendranath et al. (2001), and approaches integrating HPLC, a_{ph}, and R_{rs} by Varunan and Shanmugam (2015). These models are widely applied to satellite ocean color data, enhancing our understanding of phytoplankton community structure. More recently, Biogeochemical Argo (BGC-Argo) profiling floats have enabled large-scale assessments of carbon export from the surface to the mesopelagic zone by combining backscattering data with POC and PSCs models (Jayaram et al., 2021; Neukermans et al., 2023).

While PSCs provide a structural classification based on cell size, phytoplankton functional types (PFTs) offer a more ecologically meaningful framework by grouping species according to their roles and functions in the ecosystem. The PFTs are groupings based on ecological roles, such as nitrogen fixation, calcification, dimethyl sulfide (DMS) production, and silica uptake, rather than taxonomy, and are often linked to specific phytoplankton classes (pico, nano, and microphytoplankton). These classifications, which include groups like *Trichodesmium* (nitrogen fixers), coccolithophores (calcifiers), *Phaeocystis* (DMS producers), and diatoms (silicifiers), are essential for understanding carbon cycling and ecosystem functioning, especially under changing climate conditions (Le Quéré et al., 2005; Hood et al., 2006; Follows et al., 2007). The PFTs exhibit distinct bio-optical properties driven by differences in pigment composition and cell structure, which influence both light absorption and backscattering (IOCCG, 2014; Bricaud & Stramski, 1990). For example, absorption peaks near 440 nm and 675 nm vary depending on pigment packaging effects and phytoplankton taxa. Coccolithophores contribute to carbon sequestration through both photosynthesis and calcification and exhibit high backscattering due to their external calcite plates (Stramski et al., 2001; Loisel et al., 2007). The ratio of particulate inorganic carbon (PIC) to particulate organic carbon (POC) serves as a useful proxy to assess coccolithophore growth rates (Krumhardt et al., 2017).

Diagnostic pigments identified using HPLC, such as fucoxanthin (diatoms), peridinin (dinoflagellates), alloxanthin (cryptophytes), 19′-hexanoyloxyfucoxanthin (prymnesiophytes), 19′-butanoyloxyfucoxanthin (chrysophytes), and violaxanthin and chlorophyll-b (chlorophytes), serve as markers for PFTs (Vidussi et al., 2001; Barocio-Leon et al., 2006; Uitz et al., 2006; Wright & Jeffrey, 2006; Roy et al., 2015). These bio-optical characteristics translate into variability in R_{rs}, enabling satellite-based retrievals of PFTs using spectral decomposition and inversion algorithms (Lee et al., 2002; Bracher et al., 2009). Developing tools like hyperspectral sensors (e.g., NASA PACE) and machine learning techniques enhance the capacity to monitor shifts in phytoplankton communities and assess their biogeochemical implications on a global scale (Sadeghi et al., 2012; Xi et al., 2020).

1.3 CDOM Dynamics

CDOM is an optically active fraction of the total DOM and can be detected by remote sensing satellites (Coble, 2007; Pandi et al., 2022). It shows stronger absorption in UV and weakly in visible region of spectrum and follows an exponential decay: $a(\lambda) = a(\lambda_0)e^{(-S(\lambda-\lambda_0))}$ where $a(\lambda)$ is the absorption coefficient at wavelength λ, $a(\lambda_0)$ is the reference absorption, and S is the spectral slope indicating CDOM composition. CDOM absorption properties help trace freshwater inputs, molecular weight characterization, organic pollutants, and water masses (Coble, 2007). Its conservative or non-conservative behavior indicates source origins and transformation processes (Stedmon et al., 2003; Pandi et al., 2021b).

Fluorescence-based methods are more sensitive than absorption for assessing water quality in aquatic systems (Lakowicz, 2006; Chen et al., 2007). Protein-like fluorophores (T and B), often from phytoplankton, are effective indicators of coastal pollution, responding more rapidly than conventional parameters like BOD and ammonium. In contrast, humic-like components typically indicate terrestrial or non-pollutant sources (Murphy et al., 2008; Hur & Cho, 2012). Multivariate analysis of excitation-emission matrices (EEMs), including PARAFAC, enables calculation of indices such as FI, BIX, HIX, and RFE, which help distinguish natural vs. anthropogenic inputs and biological vs. photochemical processes (Hansen et al., 2016). Despite their potential, fluorescence techniques remain underutilized in DOM studies across the BoB and AS.

1.4 Remote Sensing and Bio-Optical Properties

In ocean color remote sensing, chlorophyll-a (Chl-*a*), a precursor pigment in phytoplankton is widely used as a proxy for estimating standing stock and primary production (PP). Nearly 60 empirical algorithms have been developed for Chl-*a* estimation, including widely used multi-band ratio models (e.g., OC2 to OC6) (O'Reilly & Werdell, 2019). Many primary production models, such as VGPM and CBPM, use Chl-*a* as a core input. Additionally, Chl-*a* concentration correlates well with other bio-optical and biogeochemical parameters, including phytoplankton absorption (a_{ph}), particulate backscattering (b_{bp}), particulate organic carbon, beam attenuation coefficient (c), and K_d (Balch et al., 2001; Xing et al., 2014; Slade & Boss, 2015; Pandi et al., 2022; Stramski et al., 2022).

Beyond Chl-*a*, empirical and semi-analytical algorithms have also been developed to retrieve other optically active substances (OAS) such as CDOM, detritus, and suspended particles (IOCCG, 2000). Key parameters like R_{rs}, absorption coefficients (a_{ph}, a_{dg}), and b_{bp} have been extensively studied for their relationships with Chl-*a*, PSCs and PFTs, and optical water types.

The variability in R_{rs} spectra with Chl-*a* concentration has also been well documented (IOCCG 3, 2000). At low Chl-*a* levels, high reflectance occurs in the blue bands (412–443 nm), while at higher concentrations, the peak reflectance shifts toward the green region (~555 nm) (O'Reilly et al., 1998; Mobley, 2022). Reflectance spectra across the visible range (400–700 nm) serve as a basis for optical water types: six optical water types based (Moore et al., 2001) and eight water types based on K_d (Jerlov, 1976).

1.5 Importance of Bio-Optical Properties in the North Indian Ocean

The North Indian Ocean (NIO) is a high productivity region, marked by its semi-enclosed northern boundaries and the absence of pole-to-pole deepwater circulation, unlike the North Atlantic or North Pacific (Qasim, 1999). This distinct geography contributes to the development of shallow oxygen minimum zones (OMZs), typically found between 160 and 1500 m depth. These OMZs are sustained by high productivity and extensive organic matter remineralization by bacterially in the subsurface, especially in the eastern NIO, where trends indicate progressive intensification of OMZs over time (Helly & Levin, 2004; Bianchi et al., 2012; Prakash et al., 2012; Schmidtko et al., 2017). The NIO comprises two principal sub-basins the AS and the BoB, each exhibiting distinct hydrographic and optical characteristics (Shetye et al., 1996; Prasanna Kumar et al., 2004; Vinayachandran et al., 2021). In this context, bio-optics provides a powerful tools for understanding regional ecosystem processes in a synoptic scale. It helps in deciphering biological productivity patterns, carbon cycling mechanisms, and light acclimation strategies of phytoplankton (IOCCG 2000; Chauhan et al., 2002a, 2002b; Chauhan and Raman, 2017; Tilstone et al., 2015). Furthermore, bio-optical tools are invaluable for detecting shifts in phytoplankton community structure driven by climate change, particularly changes in underwater light regimes and surface warming (Behrenfeld et al., 2006; Brewin et al., 2010; Bopp et al., 2013). By integrating satellite remote sensing with in situ optical observations, researchers can achieve a complete view of ecosystem dynamics in the NIO, which is critical for assessing the impacts of deoxygenation, biodiversity shifts, and biogeochemical cycling under ongoing climatic perturbations (Gomes et al., 2016; Lévy et al., 2022; Pandi et al., 2021c).

1.6 Regional Bio-Optical Studies

Building on the importance of bio-optics in the NIO, numerous regional studies have enhanced our understanding of ocean optics and biogeochemistry. Key efforts include the development of empirical algorithms for Chl-*a* retrieval (Chauhan et al., 2002a, 2002b; Nagamani et al., 2008, 2017, 2013; Tilstone et al., 2011, 2015, Preethi-Latha et al., 2014; Lotliker et al., 2018), optical water type classification using K_d and R_{rs} (Sarangi et al., 2002; Ashraf et al., 2013; Minu et al., 2016; Pandi et al., 2017), and detection of algal blooms such as *Noctiluca scintillans* and

Trichodesmium using spectral methods (Shanmugam, 2011a, 2011b; Sahay et al., 2014; Sarma et al., 2022; Dias et al., 2020b; Pandi et al., 2020). These studies also assessed bloom impacts on CDOM and microbial dynamics (Basu et al., 2011; Pandi et al., 2020; Dias et al., 2021; Dias & Kurian 2022; Chari et al., 2025).

Research on the partitioning of light absorption among phytoplankton, CDOM, and detritus has improved estimates of primary productivity (Coble et al., 1998; D'sa and Miller 2005; Del Castillo & Coble 2000; Menon et al., 2005, 2006; Pandi et al., 2014; Shanmugam et al., 2016; Vishnu et al., 2018; Sunanda et al., 2023). Algorithms for phytoplankton size classes (Sathyendranath et al., 2001; Jyothibabu et al., 2015; Sahay et al., 2017; Miranda et al., 2021; Albin et al., 2022; Krishna et al., 2024), CDOM, DOC, and POC retrieval (Shanmugam et al., 2016; Pandi et al., 2014; Shaju et al., 2019; Sanyal et al., 2020), and b_{bp} properties (Balch et al., 2001; Tiwari & Shanmugam, 2012, 2013; Preethi-Latha et al., 2013; Pandi et al., 2017) have expanded regional bio-optical applications.

Recent work on CDOM absorption and fluorescence (Chari et al., 2012, 2013, 2019, 2025; Chari and Pandi, 2017; Shanmugam et al., 2016; Sarma et al., 2018; Sanyal et al., 2020; Das et al., 2016, 2017a, b; Das & Hazra, 2019; Pandi et al., 2014, 2017, 2021a, 2021b Niloy et al., 2023) underscores its role in carbon cycling and water quality. Advances in hyperspectral sensing and satellite observations continue to enhance monitoring capabilities in the NIO.

2 Sampling and Methodology

2.1 Study Region and Sample Collection

Water samples were collected during various scientific cruises using various research vessels functioned by the Ministry of Earth Sciences (MoES), Government of India (Fig. 3.1). Those major vessels are ORV Sagar Nidhi, ORV Sagar Sampada, CRV Sagar Paschimi, and CRV Sagar Purvi were allocated for parts of the field program. These vessels are scheduled and maintained by the Vessel Management Cell (VMC) of the National Institute of Ocean Technology (NIOT). For consistent time-series data in coastal regions, vessels were also hired from M/s Sindhu Marine, Kakinada, which operates trawlers modified with onboard scientific instrumentation and laboratory setups.

2.2 Sample Collected in Cruises Information and Projects Information

For Chl-*a*, 1 L seawater sample was collected using 10 L Niskin sampler and filtered through GF/F filter (Whatman, 0.7μm) under low vacuum and partial light. The filters were extracted with 10 ml DMF (N,N-dimethylformamide) and kept it in the

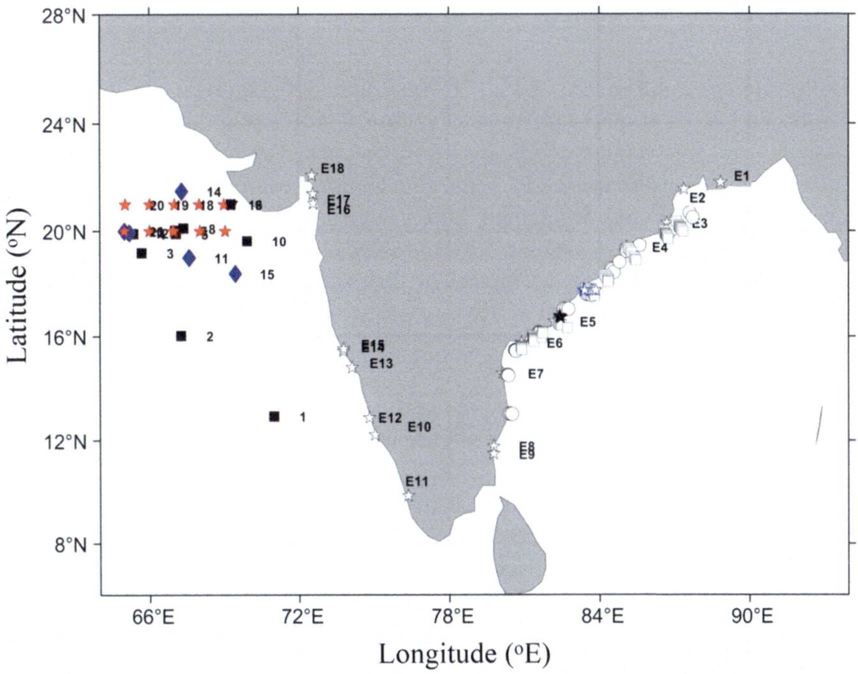

Fig. 3.1 The study region encompassed a vast range of marine environments from the AS to the BOB along India's coast. Samples were collected during scientific cruises, including the Sagarsampada cruise (Stations 1–25, three different symbols indicating three cruises) in the AS and extensive sampling in Indian estuaries (E1–E18), providing insights into river-sea ecosystems. Various research vessels like Sagar Nidhi, Sagar Sampada, Sagar Paschimi, and Sagar Purvi, along with local fishing vessels, aided sample collection, emphasizing collaborative efforts and traditional knowledge integration

dark for 24 h at 4 °C and the samples were measured fluorometrically Ex/Em: 435/670 nm following the methods of Suzuki and Ishimaru (1990). The pH was measured using bench top pH meter (Orion-2 Star). For the nutrient analysis, the GF/F filtered sea water were used and quantified their concentrations spectrophotometrically (Shimadzu UV-1800) using the standard protocols (Grasshof et al., 1999), Dissolved oxygen (DO) samples were collected using tygon tube without entering any air bubbles in 125 mL glass bottles and estimated using the modified Winkler method (Carritt & Carpenter, 1966). For CDOM absorbance and fluorescence analysis, 100 mL samples were collected onboard in pre-cleaned amber glass bottles, kept chilled at 4 °C, and transported to the lab within 6 h. CDOM fluorescence was measured using a Horiba Fluoromax-4 Spectrofluorometer with 1 cm quartz cuvettes, and instrument calibration followed Murphy et al. (2008). For phytoplankton optical properties, 0.5–1 L sea water sample was filtered onto 25-mm GF/F (Whatman, 0.7μm) under gentle vacuum and kept it for −80 °C for further analysis.

2.3 CDOM Absorbance Measurements

CDOM samples equilibrated at 24 ± 1 °C were filtered (0.22μm) and analyzed using a Shimadzu UV-2600 spectrophotometer (200–800 nm, 1 nm intervals) with 10 cm quartz cuvettes. Absorption coefficients were calculated as $a(\lambda) = 2.303 \times A(\lambda)/L$, where $A(\lambda)$ is the absorbance and L is the path length. Absorbance spectra were corrected at 700 nm (Fujiki & Taguchi, 2002), with reference wavelengths 350 and 440 nm used for CDOM estimation. Spectral slope (S), slope ratio (S_R), and $SUVA_{254}$ were derived to assess CDOM molecular characteristics and DOM aromaticity (Helms et al., 2008; Sarma et al., 2018).

2.4 Estimation of CDOM Fluorescence

Excitation-Emission Matrix (EEM) spectra were collected with the Fluoromax-4 spectrofluorometer using excitation from 250 to 402 nm and emission from 250 to 550 nm using 1 cm path length cuvette. Spectral corrections were applied using manufacturer-provided factors and normalized against the Raman peak (Ex: 350 nm, Em: 382–428 nm) following Murphy et al. (2010). Data were reported in Raman Units (RU). To identify distinct fluorophore components within the EEM spectra, Parallel Factor Analysis (PARAFAC) was applied. This multivariate technique isolates independent fluorescence components that represent chemically or functionally similar groups within the dataset. The study was conducted using MATLAB R2019a in combination with the toolbox (DOMFluor), which incorporates the N-way toolbox version 3.1 (Stedmon & Bro, 2008). Prior to modeling, regions of high-intensity signals caused by second-order Rayleigh scattering were removed by assigning zero values to affected areas, following standard correction methods. The components to retain in the model was assessed using split-half validation and random initialization techniques embedded within the toolbox. PARAFAC decomposed the dataset into five primary components (C1 to C5), each corresponding to well-known fluorescent groups: Humic fluorophores (A, C, M) and protein-like fluorophores (T and B). For each identified component, relative fluorescence intensity (I_n) was calculated as:

$$I_n = \text{score}_n \times \text{Ex}_n\left(\lambda_{\text{max}}\right) \times \text{Em}_n\left(\lambda_{\text{max}}\right)$$

where score_n is the component's relative contribution, $\text{Ex}_n(\lambda_{max})$ is the excitation loading at the excitation peak, and $\text{Em}_n(\lambda_{max})$ is the emission loading at the emission peak. This method enables semi-quantitative comparison of component intensities across samples. The fluorescence indices humification index (HIX) was calculated as the ratio of emission intensities at 435–480 nm to 300–345 nm for excitation at 254 nm. Biological index (BIX) was computed as the emission ratio at 380 nm to 430 nm at 310 nm excitation, indicating the relative abundance of protein-like fluorophores (Chiranjeevulu et al., 2014).

2.5 Phytoplankton Particulate Absorption

Particulate absorption spectra, $a_p(\lambda)$, were measured using the transmittance-reflectance (T-R) method from 350 to 750 nm at 1 nm intervals using a Shimadzu UV-2600 spectrophotometer equipped with an integrating sphere (Tassan & Ferrari, 1995a, 1995b), path amplification corrections were made using the expression given by Tassan and Ferrari (1995a). The absorption coefficient of filters was measured ($a(\lambda) = 2.303 \times OD_f(\lambda) \times S/V$, where OD_f is the optical densities of the particulate matter, V is volume in L, and S is the area of particulate matter) (Kishino et al. (1985). The phytoplankton particulate ($a_p(\lambda)$) and detritus particles ($a_d(\lambda)$) was determined before and after pigment bleaching with sodium hypochlorite, respectively. From the measured $a_p(\lambda)$ and $a_d(\lambda)$ values, the phytoplankton absorption coefficient ($a_{ph}(\lambda)$) as $a_{ph}(\lambda) = a_p(\lambda) - a_d\lambda)$, and the Chl-$a$-specific phytoplankton absorption coefficients ($a^*_{ph}(\lambda)$) were quantified by normalizing a_{ph} with Chl-a.

2.6 Estimation of R_{rs}, K_d, Chl-a

A hyperspectral underwater radiometer (Satlantic) was used to measure in situ radiometric data. R_{rs} was computed as $R_{rs} = L_w(\lambda)/E_s(\lambda)$, where L_w is upwelling radiance and E_s is downwelling irradiance (Pandi et al., 2017, 2022). Absorption (a) and b_{bp} were derived from R_{rs} using Lee et al. (2002), and the b_{bp} spectral slope (γ) was estimated via a power-law fit (Vantrepotte et al., 2011). The in situ $K_d(490)$ was derived from radiometer profiles of $E_d(\lambda, z)$ near the surface (Lee et al., 2005), and also estimated using empirical/semi-analytical algorithms for MODIS, SeaWiFS, and CZCS, then validated against in situ data (Zhang & Fell, 2007; Pandi et al., 2022). Chl-a was estimated using OC2–OC6 band ratio algorithms based on $R_{rs}(\lambda_{blue})/R_{rs}(\lambda_{green})$, with coefficients from O'Reilly and Werdell (2019). These polynomial models are standard for satellite-based Chl-a retrieval in global oceans.

2.7 Estimation of Phytoplankton Pigments

To analyze phytoplankton pigments, 3 L of surface seawater were filtered through 47 mm Whatman GF/F filters (0.7μm pore size) under low-light conditions to prevent pigment degradation. The filters were immediately stored at −80 °C for later HPLC analysis. For extraction, each filter was placed in 3 mL of acetone (95%), subjected to ultrasonication in an ice bath for 5 min, and stored at −20 °C overnight. Additional information on the HPLC procedure and calibration standards are detailed in Pandi et al. (2014, 2022). Phytoplankton size classes fractions (pico, nano, and micro) were quantified using diagnostic pigments (Babin et al., 2003; Uitz et al., 2006). The size fractionated percent contribution of phytoplankton were calculated using:

$$fp(\%) = \frac{100\,\mathrm{x}\left(0.86[\mathrm{zeaxanthin}] + 1.01[\mathrm{Chl\,b}]\right)}{\mathrm{DP}}$$

$$fn(\%) = \frac{100\,\mathrm{x}\left(0.60[\mathrm{alloxanthin}] + 0.35[19' - \mathrm{BF}] + 1.27[19' - \mathrm{HF}]\right)]}{\mathrm{DP}}$$

$$fm(\%) = \frac{100\,\mathrm{x}\,(1.41[\mathrm{fucoxanthin}] + 1.41[\mathrm{peridinin}]}{\mathrm{DP}}$$

The diagnostic pigment (DP) index, photoprotective pigments (PPP), photosynthetic accessory pigments (PSP), and the phytoplankton size index (SI) were quantified using the equations of Bricaud et al. (2004). The abbreviations and terms used in this study are listed in Table 3.1, and the equations for calculating error statistics to validate empirical and algorithm-based results are provided in Table 3.2.

Table 3.1 Abbreviations and terms commonly used in the present study

Abbreviation	Description	Unit
$a_{CDOM}(\lambda)$	Absorption by Coloured Dissolved Organic Matter (CDOM)	m^{-1}
Chl-*a*	Chlorophyll	mg m^{-3}
$a_{ph}(\lambda)$	Phytoplankton absorption	m^{-1}
$a^*_{ph}(\lambda)$	Chl-*a* specific absorption by phytoplankton	m^2 mg^{-1}
$a_w(\lambda)$	Absorption of pure water	m^{-1}
$a_{dg}(\lambda)$	Absorption of detritus plus CDOM	m^{-1}
$R_{rs}(\lambda)$	Remote sensing reflectance ($R_{rs} = L_u/E_d$) L_u = upwelling radiance, E_d = downwelling irradiance	sr^{-1}
$E_d(\lambda)$	Downwelling spectral irradiance	mW cm^{-2} nm^{-1}
$K_d(490)$	Diffuse attenuation coefficient at 490 nm	m^{-1}
$nL_{wn}(\lambda)$	Normalized water-leaving radiance	mW cm^{-2} nm^{-1} sr^{-1}
$L_w(\lambda)$	Water-leaving radiance	mW cm^{-2} nm^{-1} sr^{-1}
PAR	Photosynthetically active radiation	μmol photons m^{-2} s^{-1}
Z_{eu}	Depth of the euphotic layer, defined as the depth where the PAR is reduced to 1% of its surface value	m
DCM	Deep chlorophyll maximum	m
MLD	Mixed layer depth	m
K_d(PAR)	Diffuse attenuation coefficient of PAR	
DAM	Deep photoAcclimation maxima	m
DCM	Deep chlorophyll maximum (DCM)	m
PSP	Photosynthetic pigments	μg/L
PPP	Photoprotective pigments	μg/L
OSS/OAS	Optically significant substances/optical active substances/	m^{-1}
C_{pd}	Chlorophyll mean upto penetration depth/first optical depth	mg m^{-3}
C_t	Column integrated chlorophyll	mg m^{-2}

Table 3.2 Equations used for calculating error statistics for validation of algorithms

Statistical estimator	Formula
Bias	$\frac{1}{n}\sum_{n}^{i=1}(y_i - x_i)$
Scatter-index	$\frac{1}{\bar{x}}\sqrt{\frac{\sum_{n}^{i=1}[(y_i - \bar{y}) - (x_i - \bar{x})]^2}{}}$
Root mean square error (RMSE)	$\sqrt{\frac{1}{n}\sum_{n}^{i=1}(y_i - x_i)^2}$

3 Bio-Optical Differences in the BOB and the AS

The AS stands out as a highly upwelling and productive region characterized by minimal terrestrial influence from major rivers. Its waters predominantly exhibit characteristics of Case 1 waters, with phytoplankton and related substances comprising the primary OAS (Table 3.3). This simplicity in optical properties facilitates ocean color remote sensing with fewer complexities. In contrast, the BOB experiences significant river runoff from the coastal rivers of East India, leading to a high influx of suspended matter, particularly during monsoon seasons. This suspended matter forms a stratified layer at the surface, hindering light penetration and subsequently reducing productivity. The BOB also displays heightened optical complexity, primarily due to the interaction between various OAS, which are largely Case 2 in nature. Furthermore, the BOB's coastal regions is more prone to cyclones than the AS, which bring about physical disturbances that can disrupt the natural ecosystem for several months. Additionally, the presence of eddies further influences the coastal dynamics of the BOB. While algal blooms in the AS are recurrent and some can be harmful, those in the BOB are comparatively less harmful due to the enrichment of silicate sources from terrestrial runoff. This disparity in optical and ecological characteristics between the AS and the BOB underscores the diverse and dynamic nature of the Indian Ocean's coastal regions.

Table 3.3 CDOM absorption coefficients, spectral slopes, and conservative (C) and non-conservative (NC) behavior reported in various coastal regions of the world

Name of estuary	Salinity (psu)	λ_{nm}	$a_{CDOM}(\lambda)$ (m^{-1})	Wavelength (nm)	Slope (nm^{-1})	CDOM Behavior	Reference
Amazon	15.1–36.4	355	0.14–3.12	300–500	0.015–0.033	NA	Green and Blough (1994)
St Lawrence	0–35.5	365	0.2–3.8	300–500	NA	C	Nieke et al. (1997)
Orinoco River plume, Eastern Caribbean Sea	19.7–33.7	355	0.43–4.17	300–450	0.012–0.020	C	Del Castillo et al. (1999)
Chesapeake Bay	13–33 (April)	355	0.5–2.2	280–650	0.0160–0.0195	C	Rochelle-Newall et al. (2002)
	5.8–27 (October)	355	0.7–2.1			C	
Baltic Sea	NA	440	0.005–20	300–500	0.0193		Schwarz et al. (2002)
Narragansett Bay	NA	412	0.09–1.4	400–550	0.006–0.068	C	Keith et al. (2002)
Southern Atlantic Bight (Cape Fear River and Onslow Bay)	23–37	350	1.4–12.3	300–500	0.0154–0.0156	C	Kowalczuk et al. (2005)
Pearl	0–34.96	355	0.34–1.4	NA	NA	NC	Chen et al. (2004)
Pearl	0–33	355	0.24–1.93	300–500	0.0138–0.0184	C	Hong et al. (2005)
Funka Bay		440	0.022–0.14	370–440	0.011–0.024		Sasaki et al. (2005)
Tampa Bay	1.1–32.5	350	1.14–30.21	350–440	0.013–0.024	NC	Chen et al. (2007)
Tampa Bay	NA	400	1.11–7.76	350–440	0.013–0.024		Chen et al. (2007)
Pearl River estuary	NA	443	0.09–0.35	400–500	0.015–0.0185	–	Zhao et al. (2009)
Taiwanese estuaries	0–33.9	350	0.15–12.5	350–400	0.0152	C	Yang et al. (2013)
Southern Beaufort Sea	NA	440	0.018–1.08	350–500	0.015–0.023	–	Matsuoka et al. (2011)
Malaysian coast facing South China Sea	NA	440	0.01–0.43	350–600	0.0082–0.026 (0.016)	–	Bowers et al. (2012)
Coastal BOB (laboratory cultures)	33	350	2.1–8.5	300–440	0.011–0.033	–	Chari et al. (2013)
Western BOB	20–34	440	0.021–0.25	300–500	0.009–0.022	C, NC	Pandi et al. (2014)
Coastal BOB	11.67–34.95	350	0.322–5.067	300–500	0.009–0.0146	C	Chiranjeevulu et al. (2014)

Hugli Estuary, West Bengal coast, BOB	1–33	440	0.086–0.24	400–700	0.0071–0.0167	C	Das et al. (2016)
Western BOB	20–35	440	0.01–0.45	–	–	NA	Lotliker et al. (2020)
Rushikulya estuary, BOB		440	0.02–4.48	400–700	0.001–0.015	–	Parida et al. (2019)
Northern BOB	16–33	440	0.10–0.70	400–700	0.0071–0.0229	C	Das et al. (2017a)
Coastal, northern BOB	16–33	440	0.11–0.142	400–700	0.0095–0.0141	C	Das et al. (2017b)
Godavari estuary, Coastal BOB	1–33	350	0.64–5.76	300–500	0.0157, 0.011–0.016	C, NC	Pandi et al. (2021a)
Coastal BOB, off Visakhapatnam	22–34	440	0.031–1.43	300–500	0.007–0.020	C, NC	Pandi et al. (2021b)
Godavari estuary	0.1–35.2	350	0.64–5.76	300–500	0.011–0.016	C & NC	Pandi et al. (2021a) and Chari et al. (2019)
Sundarban, BOB	18–25	350	2.5–7.6	290–500	0.014–0.022	C	Saynal et al. (2020)
All Indian estuaries	0–29	350	0.59–4.78	300–500	0.011–0.0172		Sarma et al. (2023b)
Coastal BOB, off Yanam	1.32–34.82	350	0.29–3.22	300–500	0.007–0.017	C	Pandi et al. (2025)
West coast							
Arabian Sea (AS)	35–36.4	375	0.02–0.29	200–750	NA	NC	Coble et al. (1998)
AS	NA	375, 350	0.02–0.33	350–650	0.012–0.024	–	Breves and Reuter (2000)
Northeastern AS	NA	440	0–12	–	–	C	Menon et al. (2006)
Mandovi and Zuari estuaries, Lakshadweep waters, AS	NA	440	0.22–21.66	400–650	–	–	Menon et al. (2005)
Eastern AS	NA	325	0.28–2.27	300–700	0.0168–0.0385		Basu et al. (2011)
SW coast of India, AS, off Cochin	25–35	443	0.066–0.185	400–650	0.012–0.014	C	Vishnu et al. (2018)
Arabian Sea, coastal waters, off Goa	35–36	412	0.092–32.23	–	–	–	Dias et al. (2020b)
Northeastern AS	33–36	440	0.115–0.276	–	–	–	Pandi et al. (2020)
Mandovi estuary, AS	22–35		0.179–0.203	275–295	0.0233 0.0240	C	Dias et al. (2021)
Eastern AS	–	440	0.37–0.50	400–650	–	–	Patil et al. (2023)
AS, off Ratnagiri	NA	440	0.08–0.30	400–550	0.0076–0.0126	NA	Krishna et al. (2024)

4 Bio-Optical Studies in BOB

4.1 IOPs and AOPs Variability in the BOB

The main monsoon (summer, southwest monsoon) season begins in June in the northern Indian Ocean and lasts till September–October. During the pre-monsoon period from January to May, optical properties were measured off Visakhapatnam's coastal region in the BOB (Pandi et al., 2017). Notably, it was found that anomalies in particle b_{bp} and R_{rs} induced by a supercyclone (Hudhud, October 2014) persisted for 5 months up to the 50 m isobath/15 km from the coast and 6 months up to the 30 m isobath/5 km away. These anomalies (higher than usual b_{bp} and R_{rs}) were attributed to fine suspended particles leaching from coastal deposits formed during the cyclone. The R_{rs} spectra exhibited four distinct types (Fig. 3.2), including Type 1 with a reflectance hump (at ~475 nm) attributed to fluorescent CDOM and Types 2, 3, and 4 corresponding to different proportions of chlorophyll, CDOM, and total suspended matter (TSM). The b_{bp} spectra showed regular decreases from shorter to longer wavelengths. Anomalies included significantly higher b_{bp} and smaller γ values, indicating the presence of small, highly reflecting particles post-cyclone. Additionally, $R_{rs}(685)$ exhibited abnormal values, rendering it unsuitable for Chl-a retrieval during this period. The anomalous suspension likely originated from coal-like debris and plastic fragments deposited by the supercyclone near Visakhapatnam harbor.

In the western BoB, Pandi et al. (2014) observed a significant inverse relationship between salinity and CDOM absorption at 440 nm ($a_{CDOM}(440)$), suggesting conservative mixing between terrestrial and marine waters. The spectral slope ($S_{300\text{–}500}$) averaged 0.0141 ± 0.002 nm^{-1} and correlated inversely with $a_{CDOM}(440)$,

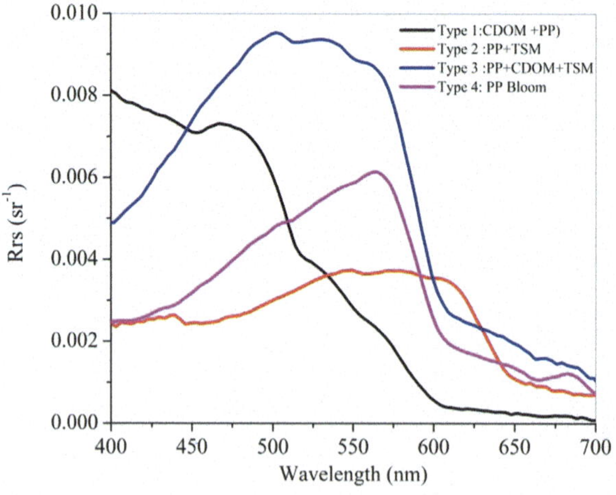

Fig. 3.2 Typical examples of remote sensing reflectance (R_{rs}) spectra showing types 1–4 (3 February 2014)

aligning with global averages. In northern transects, lower slope values (mean $S = 0.0121\ nm^{-1}$) indicated photochemical CDOM degradation in frontal zones formed by the convergence of Ganga and peninsular river discharges. Das et al. (2017a, b) further noted seasonal variation in spectral slopes, with lower values during the monsoon ($0.0123\ nm^{-1}$) compared to non-monsoon periods ($0.0185\ nm^{-1}$). Similar patterns were reported by Parida et al. (2019) off the Odisha coast. Ganga water, originating from the high-elevation Himalayas (3892 m), is considered younger and more labile than the more humified DOM from peninsular rivers. The spectral slope ratio (S_R) was conservative for samples with salinity <32 PSU. CDOM molecular weight was higher in samples diluted by Ganga water than local sources, indicating larger, more humified DOM inputs. The northern BoB (nBoB) is dominated by microplankton and high CDOM contributions to total absorption (aT), with CDOM and phytoplankton accounting for ~41% and 20%, respectively, during the monsoon. Moreover, Das et al. (2016) first observed a significant linear relationship between salinity and $a_{CDOM}(440)$ of the Ganga estuary water up to a salinity of 22 PSU. Therefore, $a_{CDOM}(440)$, which was mostly allochthonous in the region, could be considered a proxy of salinity with acceptable certainty (Das et al., 2016, 2017a, b). CDOM values also positively correlated with the mean monthly rainfall of this estuarine region (Das et al., 2017a, b). Das et al. (2019) have observed that the CDOM absorbance decreased gradually from the Ganga River confluence region to the offshore region.

The tropical River Godavari, India's second-longest after the Ganga and 32nd globally, undergoes heavy monsoon flooding and experiences tidal incursion during the non-monsoon (summer) season (Pandi et al., 2021b). In summer, high temperatures allow warm, saline coastal water to intrude into the lower estuary. Between April 2010 and March 2011, 18 field campaigns measured CDOM, with $a_{CDOM}(350)$ ranging from 0.645 to 5.759 m^{-1}, revealing eight optically distinct water types. CDOM from four sources during low-to-moderate discharge (salinity <33) behaved conservatively. At salinity >33, higher CDOM suggested a negative estuary condition driven by summer heat and coastal currents. The upper estuary during nil discharge showed signs of photochemical and bacterial CDOM degradation. During peak discharge (90–92% of annual flow), CDOM behaved non-conservatively and showed a strong inverse relation to pH, predicting ~1.2 m^{-1} $a_{CDOM}(350)$ export to the coastal sea at marine pH (8–8.2). Remarkably, values exceeded 3.29 m^{-1} during the first summer flood. CDOM was mostly unaffected by anthropogenic inputs, except occasionally during the dry season (Pandi et al., 2021b, 2021c). Analysis of over 800 absorption spectra from the Godavari estuary and Chilika lagoon showed that the spectral slope $S_{330-440}$ best captures the exponential nature and variability of CDOM (Sarma et al., 2018).

Seasonal variation of fluorescent dissolved organic matter (FDOM) in the Godavari estuary and adjacent BOB coastal waters revealed a shift in the UV humic to visible humic-like fluorophore ratio—lower in the pre-monsoon and higher during the monsoon. This suggests the pre-monsoon is bacteria-rich, leading to in situ CDOM degradation, while the monsoon is bacteria-poor, with terrestrial CDOM remaining largely intact and only photodegraded. Tyrosine protein-like fluorophore indicated bacterial degradation during pre-monsoon (Chari et al., 2012). The southwest monsoon brought terrestrial organic matter (UV, visible, and marine humic-like) into northern coastal BOB, while autochthonous production marked the southern region, also shown by tyrosine protein-like fluorophore (Chari et al., 2013). Eddies, common in eastern BOB, may result from the northward East India Coastal Current (EICC), bordered by offshore warm and inshore upwelled cool waters (Shetye et al., 1993; Rakesh et al. 2008). A cold-core eddy observed in northern coastal BOB during summer showed enriched bacterially degraded organic matter, indicated by elevated tyrosine protein-like fluorophore (Chiranjeevulu et al., 2014). This fluorophore can be masked by tryptophan-like fluorophore when phytoplankton-derived organic matter is high, e.g., in summer estuaries, supported by a positive DOC correlation in the Godavari estuary (Chari et al., 2019). Diurnal optical property data indicated that lower molecular weight, aromatic compounds form more during summer than winter (Chari et al., 2019). Humic-like matter formed during pre-monsoon from phytoplankton excretion in coastal waters, confirmed via axenic lab studies on dominant BOB phytoplankton species (Chari et al., 2013).

4.2 Assessment of Phyto Size Class Models in the BOB

In their study, Miranda et al. (2021) utilized the algorithms developed by Sahay et al. (2017) to quantify different phytoplankton size classes in the coastal waters of the BOB, specifically off the regions of Gopalpur and Visakhapatnam. Their analysis revealed promising performance of both models, accurately retrieving nano and microphytoplankton populations. However, in the case of picophytoplankton, the model proposed by Brewin et al. (2012a) performed poorly compared to Sahay et al. (2017) model. To delve deeper into the findings, Sahay et al.'s model demonstrated superiority in the BOB context, particularly in accurately capturing picophytoplankton dynamics. Leveraging this model, Miranda et al. (2021) extended their investigation by applying it to MODIS-Aqua Chl-*a* data. This expanded analysis aimed to uncover temporal and spatial variations in phytoplankton distribution, ranging from nearshore to offshore regions. The results illuminated a distinct spatial pattern in phytoplankton composition across the study area. Nearshore regions exhibited a dominance of microphytoplankton populations, indicating favorable conditions for their proliferation closer to the coast. In contrast, offshore areas showcased a prevalence of picophytoplankton, suggesting environmental factors conducive to their thriving further from the shore.

4.3 Assessment of Empirical Algorithms of Chl-a

Nagamani et al. (2013) validated various Chl-*a* empirical algorithms in the BOB using satellite and in situ data across a 0.2–4.0 mg m^{-3} range. Chl-*a* values were retrieved from OCM-2 products using global ocean color algorithms (OC2, OC3,

OC4, and Chl-*a*). OC2 showed the strongest correlation with in situ Chl-*a* ($R = 0.61$), while other algorithms consistently overestimated Chl-*a*, even at low concentrations. The study recommended developing regional algorithms and fluorescence-based methods for improved accuracy. Singh and Chaturvedi (2010) compared IRS-P4 OCM with MODIS-Aqua Chl-*a* in the AS and BOB, finding better correlation in the AS, especially in December. Lotliker et al. (2020) studied long-term Chl-*a* and OAS (TSM, CDOM) variability in northwestern BOB, observing peak Chl-*a* pre-SW monsoon and peak OAS during the SW monsoon (Aug–Oct). Satellite-derived Chl-*a* (MODIS-Aqua, OCM-2, VIIRS) showed overestimation in nearshore waters, with errors ranging from 33% to 51%; MODIS-Aqua was most accurate. Preethi-Latha et al. (2014) analyzed in situ fluorometer Chl-*a* data against OCM-2 using OC4-V4 and OC2 algorithms, noting OCM-2 overestimated at low and underestimated at high Chl-a levels. Earlier, Preethi-Latha et al. (2013) examined BOB coastal water b_{bp}, finding good agreement between in situ and quasi-analytical $b_{bp}(700)$, but a nonlinear relationship between in situ $b_{bp}(770)$ and Chl-*a* < 5 mg m^{-3}, linked to larger phytoplankton particles in nearshore vs. offshore waters.

In the BOB during the southwest monsoon, Ashraf et al. (2013) observed a well-defined linearity between the K_d at various wavelengths especially shorter wavelengths and Chl-*a*. This emphasized the significant influence of phytoplankton in causing variability in $K_d(\lambda)$ in these regions; the relationship weakened at longer wavelengths due to strong absorption by seawater. Shanmugam et al. (2013) developed CDOM-based algorithms in the southwestern BOB for dissolved organic carbon (DOC) and salinity. These algorithms were validated using in situ data collected from various domains within the BOB, e.g., inland, coastal, and oceanic regions.

5 Bio-Optical Properties in the AS

5.1 Bio-Optical Algorithms During Peak Noctiluca Blooms

In 2014, scientists from the Lamont-Doherty Earth Observatory (LDEO) opened a big debate on the possible cause of the mixotrophic NS blooms. They asserted (Gomes et al., 2014) that human activities, particularly sewage discharge and other anthropogenic factors from the Indian coast, were responsible for their occurrence and exacerbation in the affected region. Such findings underscore the importance of mitigating pollution and managing coastal development to safeguard marine ecosystems from harmful algal blooms. In 2018, scientists from the National Institute of Oceanography (NIO*) challenged this assertion by identifying the stable isotope signature of carbon (C) and nitrogen (N) in the NS blooms, which matched with that of NS bloomed under controlled conditions. The NIO* scientists went on further to show that the NS blooms were caused by the shallowing of the mixed layer due to West India Coastal Current (WICC) activity. They highlighted how this process led to nutrient depletion in the upper ocean layers, creating favorable conditions for opportunistic NS species to flourish. By elucidating the mechanisms driving NS bloom dynamics, particularly the role of natural nutrient cycling processes and oceanographic phenomena like the WICC, the NIO* study provided valuable insights into the complex interactions shaping marine ecosystems in the region. More recently, Sarma et al. (2022, 2023) have shown that the timing of the intensification of NS blooms in the NAS happening at a time when (1) the arrival of the northward flowing WICC takes place at the northeast slope and (2) the cool northeasterly winds begin to slow down and give place to the warm northwesterly winds that introduce iron into the surface water (Sarma et al., 2022). These findings underscored the importance of considering both natural and anthropogenic factors in understanding and managing harmful algal blooms and their ecological impacts (Fig. 3.3). During intense winter periods characterized by peak northeast (NE) winds, diatom blooms are stimulated due to nutrient upwelling. However, during seasonal transitions, the presence of the WICC impedes deep upwelling, resulting in local depletion of dissolved iron before silicate. Subsequently, when northwesterly winds become prominent, continental iron inputs replenish the deficit, providing a trigger for the formation of intense NSG (green Noctiluca) blooms. These

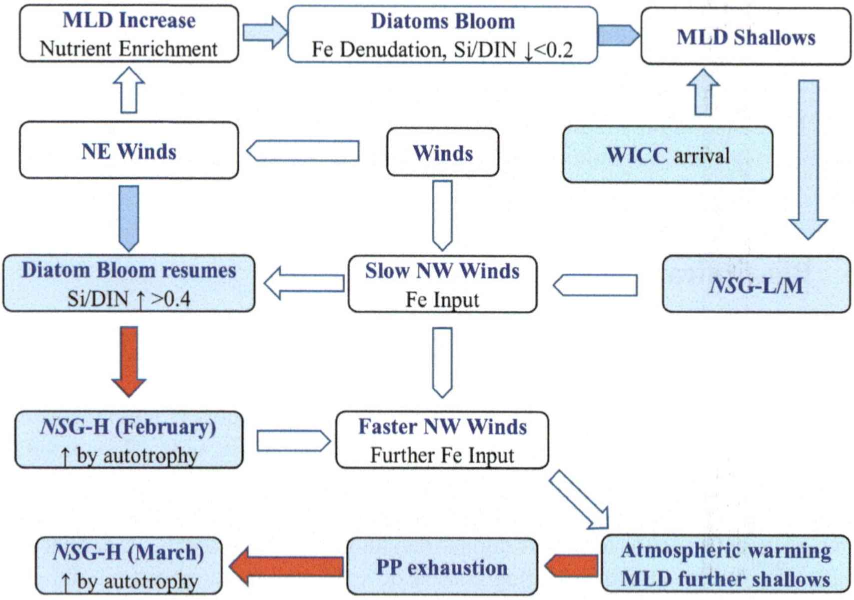

Fig. 3.3 Conceptual diagram illustrating the formation of *green* Noctiluca scintillans (NSG) blooms in the NEAS, based on findings from this study and prior research. The diagram uses color-coded boxes: unshaded for atmospheric or phytoplankton-related drivers, blue-filled for oceanographic processes, and red-filled to denote NSG bloom events. Text inside each box represents specific outcomes and their respective drivers. Key terms: *MLD* mixed layer depth, *PP* non-Noctiluca phytoplankton, *NW/NE* northwest/northeast directions, *WICC* West India Coastal Current, *DIN* dissolved inorganic nitrogen. (Adapted from Sarma et al., 2022)

blooms are fueled by an increase in the heterotrophic-dominant stock of Noctiluca, transitioning to an autotrophy-dominant stage of nutrition. The synthesis of findings from this study and previous research provides insights into the formation and intensification of NSG blooms in the northeastern AS, as depicted in the accompanying scheme.

Various research teams had earlier noted ~25 °C as the threshold SST for the formation of NS blooms. Our research team achieved a significant milestone by developing remote sensing (R_{rs}) algorithm tailored for real-time monitoring of NS (*Noctiluca scintillans*) blooms, particularly during early stages characterized by sea surface temperatures (SST) below 25 °C (Fig. 3.4). Leveraging Chl-*a* levels as a proxy, we integrated the spectral reflectance ratio $R_{rs}(531)/R_{rs}(510)$ into the model. This ratio was chosen due to the distinctive reflectance pattern exhibited by NS, with a peak at approximately 531 nm and a trough at around 510 nm, corresponding to low absorption/high reflectance and peak absorption/minimum reflectance, respectively. The developed algorithm, expressed as $\frac{Rrs(531)}{(Rrs(510)} = 0.8261 + 6.06 \times 10^{-6} \times \text{NS} + 0.02323 \times \text{Chl-}a$, provided a robust framework for real-time *NS* monitoring based on satellite-derived remote sensing data. Building upon this foundational work, our team made further advancements in *NS* bloom monitoring by proposing an enhanced remote sensing algorithm in 2023. This algorithm expanded the monitoring capability to encompass a wider range of SST conditions, a crucial consideration given the dynamic nature of NS blooms. Integrating

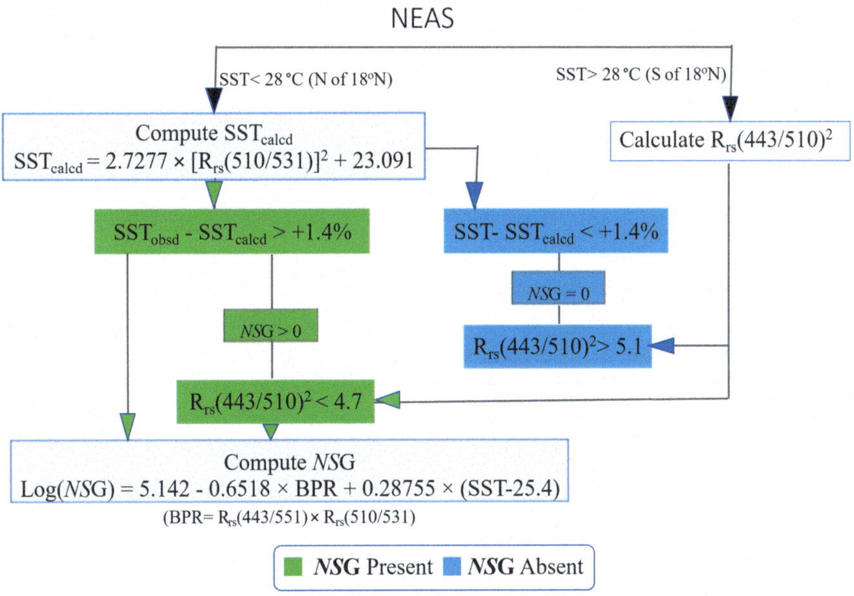

Fig. 3.4 Flow chart of detection and estimation of NSG in the northeastern AS (Fig. 9, therein Sarma et al., 2023b)

SST as an additional parameter, the algorithm was formulated as $\text{Log}(\text{NS}_{\text{green}}) = 5.142 - 0.652 \times \frac{Rrs(443)}{(Rrs(551)} \times \frac{Rrs(531)}{(Rrs(510)} + 0.288 \times (\text{SST-25.4})$. By incorporating SST into the algorithm, we provided a more comprehensive tool for NS bloom monitoring across varying environmental conditions. To facilitate the practical application of these algorithms, we developed a detailed scheme outlining the procedures for real-time monitoring of NS, focusing specifically on the green variant of NS, i.e., NS green (NSG) (Sarma et al., 2023a). This resource serves as a valuable guide for researchers, environmental agencies, and stakeholders involved in the monitoring and management of NS blooms, offering insights into the utilization of R_{rs} techniques for timely and effective surveillance of these ecologically significant events.

We further put forward a model that derives NSG abundance when both SST and (total) phytoplankton abundance are available (Sarma et al., 2023b). This model is applicable during both the heterotrophic phase, characterized by an increase in phytoplankton abundance, and the autotrophic phase, marked by a decrease in phytoplankton abundance. The proposed model is expressed as $R = 10^{-5} \times \text{NS} + 0.8$, where R represents the ratio of the logarithm of NS abundance to the logarithm of phytoplankton abundance other than NS. During the heterotrophic phase, when phytoplankton abundance increases, the model predicts that the ratio R will decrease as NS abundance rises relative to the overall phytoplankton community. Conversely, during the autotrophic phase, when phytoplankton abundance decreases, the model suggests that the ratio R will increase, indicating a higher prevalence of NS relative to other phytoplankton species. By incorporating both phases of NS dynamics and their relationship with phytoplankton abundance into a single model, this approach offers a comprehensive framework for understanding and predicting the dynamics of NS blooms in marine ecosystems.

5.2 *Coupling of CDOM and DIC in the Indian Estuaries*

In estuarine environments, the composition of chromophoric dissolved organic matter (CDOM) is predominantly characterized by fulvic acids and humic acids (FA/HA) (Sarma et al., 2023b). These compounds are akin to inorganic weak acids such as carbonic acid, boric acid, and silicic acid and play a significant role in contributing to total alkalinity within estuarine systems. A model has been developed to calculate dissolved inorganic carbon (DIC) based on CDOM concentration and pH levels. The model equation, DIC (μm) = $-63.05 + 934.67\text{pH} + 199.84\text{CDOM}$ (a_{350}, m^{-1}), incorporates CDOM absorbance at 350 nm (a_{350}) and pH values. This model provides a means to estimate DIC concentrations, critical for understanding carbon dynamics within estuarine environments. Furthermore, CDOM can serve as an indicator for predicting estuarine acidification by assessing deviations from the spectral slope ratio (S_R) with changes in salinity. The S_R, calculated as $(S_{275-295})/(S_{350-400})$, exhibits a linear relationship with salinity, expressed by the equation $S_R = 0.009*\text{Salinity} + 0.85$. This relationship aids in identifying changes in CDOM composition associated with variations in salinity, offering insights into potential acidification processes within estuarine systems. Studies have revealed that CDOM can also serve as a marker for anthropogenic organic pollution in estuaries, particularly in southeastern and northwestern Indian estuaries. Additionally, differences in continental weathering between estuaries, such as those in the northeast BOB and southwest AS regions, result in variations in pH levels. Specifically, estuaries experiencing higher levels of continental weathering exhibit pH values approximately 0.48 units higher than those in estuaries with lower continental weathering at corresponding salinity levels. These findings underscore the complex interplay between CDOM dynamics, environmental factors, including natural processes like continental weathering and rainfall, and anthropogenic influences in estuarine ecosystems.

5.3 *Optical Water Types*

Optical classification methods based on the shape and magnitude of R_{rs} spectra have been applied globally and regionally. In the Indian Ocean, Sarangi et al. (2002) classified waters off Gujarat using K_d and Chl-*a* into four types: IA (K_d: 0.035–0.040 m^{-1}, Chl-*a*: 0.05 mg m^{-3}), IB (0.042–0.065 m^{-1}, 0.1 mg m^{-3}), II (0.07–0.1 m^{-1}, 0.5 mg m^{-3}), and III (0.115–0.14 m^{-1}, 1.5–2.0 mg m^{-3}), following Jerlov's classification scheme. Minu et al. (2016) further classified western AS coastal waters (off Kochi) into three optical types based on R_{rs} spectral shape. Type-I, with low Chl-*a*, CDOM, and $b_{bp}(650)$, peaked at 470 nm, indicating pure Case-I waters. Type-II, with elevated bio-optical properties, showed peak R_{rs} around 560–570 nm. Type-III waters exhibited a further redshift in R_{rs} (570–700 nm) and a strong secondary peak at 684 nm due to higher Chl-*a* and CDOM. Six empirical algorithms (OC3C, OC4O, OC4, OC4E, OC3M, OC4O2) used with sensors like CZCS, SeaWiFS, MODIS, and OCM-2 performed best in Type-I waters but poorly in Type-II and III. The Chl-*a* and CDOM correlation suggested river discharges as major sources of phytoplankton-supporting nutrients.

5.4 *Optical Classification of Coastal Waters in the Northern Indian Ocean*

Monolisha et al. (2018) characterized the R_{rs} spectra in the AS and BOB using a 15-year climatological dataset (1998–2013) from the Ocean Color Climate Change Initiative. The study employed fuzzy logic optical classification, following the methodology of Moore et al. (2009), to identify eight distinct optical water classes across the coastal northern Indian Ocean. Classes 1–6 were associated with Case-1 waters, typical of deeper, oligotrophic environments, while Classes 7 and 8 corresponded to Case-2 waters, found in inshore eutrophic zones with high chlorophyll concentrations, sediment load from riverine inputs, and land runoff. The mean spectral signatures of these classes highlighted this distinction, demonstrating the influence of biological and physical constituents on water color. Seasonal analysis revealed significant shifts in optical class distribution during the summer and winter monsoons and the intermonsoon periods. These changes, driven by physical forces like winds, currents, rainfall, and sediment influx, highlight the dynamic coastal optical variability and its link to environmental processes in the northern Indian Ocean.

5.5 Chlorophyll-a Algorithm Evaluation in the Northern Indian Ocean

The performance of various empirical algorithms for retrieving Chl-*a* from ocean color satellite data has been extensively evaluated in the northern Indian Ocean, including the AS and BOB. These studies have focused on the effectiveness of both global and regionally tuned algorithms under different bio-optical conditions, particularly in Case-I (clear) and Case-II (coastal/turbid) waters.

Chauhan et al. (2002a) tested six empirical Chl-*a* algorithms—Aiken-C, POLDER-C, OCTS-C, Morel-3, OC2, and OC4—using AS data. Among them, OC2 and OC4 exhibited superior performance in both Case-I and Case-II waters. OC2, which uses the $R_{rs}(490)/R_{rs}(555)$ band ratio, was preferred for operational use due to its better atmospheric correction at 490 nm, compared to the OC4 algorithm that uses four bands (443, 490, 510, 555 nm). However, OC4 struggled in Trichodesmium-dominated waters, leading to inaccuracies in Chl-*a* estimation.

To enhance Chl-*a* retrieval accuracy, Nagamani et al. (2008) developed the OC4 v4 algorithm specifically for the AS, utilizing a Multi-band Ratio (MBR) approach with the same four bands and incorporating them into a fourth-order polynomial. This algorithm, designed for the OCM sensor aboard OCEANSAT-2, aimed to improve Chl-*a* estimation in regional waters.

Tilstone et al. (2011) compared OC4v6, Carder, and OC5 algorithms in both the BOB and open AS. OC5 consistently outperformed the others, showing the lowest Root Mean Square Error (RMSE), minimal bias, and a regression slope near 1. In contrast, OC4v6 and Carder overestimated Chl-*a* concentrations in the coastal BOB, likely due to elevated Total Suspended Matter (TSM), revealing their sensitivity to bio-optical variability. In a follow-up study, Tilstone et al. (2013) performed a time-series analysis in the eastern AS, comparing OC3M, the Garver–Siegel–Maritorena (GSM) model, and the Generalized Inherent Optical Properties (GIOP) algorithm. OC3M demonstrated the best performance with only 11% error, compared to 24% for GSM and 55% for GIOP. OC3M also showed resilience to errors in R_{rs} and variations in CDOM and TSM. The study noted seasonal variability, with maximum Chl-*a* concentrations during the winter monsoon on the northeastern shelf.

Sathyendranath et al. (2001) assessed the accuracy of CZCS, OCTS, and SeaWiFS algorithms for estimating Chl-*a* concentrations in the AS, ranging from 0.03 to 6 mg m^{-3}. SeaWiFS wavelengths produced the most reliable results (error < 25%), whereas OCTS wavelengths performed the worst (errors up to 120%), and CZCS fell in between (~50% error). The study also identified a strong correlation between $a_{ph}(443)$ and Chl-*a*, and classified phytoplankton into two populations based on their specific absorption coefficients. Diatoms and prymnesiophytes dominated the samples, as inferred from pigment ratios. Algorithm comparison indicated strong alignment with CZCS and SeaWiFS, while OCTS results diverged significantly.

Minu et al. (2014) explored bio-optical variability off the Kochi coast, revealing a negative correlation between $a_{CDOM}(440)$ and salinity during the southwest monsoon, suggesting conservative mixing of marine and terrestrial waters. A good correlation was observed between R_{rs}-derived Chl-a using OC3M and in situ Chl-a, but the agreement was weaker when using atmosphere-corrected A1 and A2 bands. No significant difference was noted between A1 and A2 results, likely because the region had low CDOM and TSM concentrations. Additionally, a spectral shift in Rrs toward higher wavelengths was observed with increasing Chl-*a*. Overall, these studies indicate that algorithm performance varies considerably across different bio-optical environments. OC2 and OC3M emerged as robust options for the AS, while OC5 showed promise in handling coastal BOB conditions. However, global algorithms often struggled with high TSM and CDOM levels, reinforcing the need for region-specific algorithm development and validation in the northern Indian Ocean.

5.6 *Empirical Algorithms for* K_d

Chauhan et al. (2003) developed an algorithm for estimating the K_d using radiometer profiles' data. The formula for this algorithm was: $K_d(490) = 0.022 + 0.124\ [L_{wn}(443)/L_{wn}(555)]^{-1.64}$. This algorithm was then validated using OCM-2 data products, which exhibited a highly significant relationship with in situ data. The validation results indicated a strong correlation, with an R^2 value of 0.80, based on 41 data points, and a RMSE of 0.011 m^{-1}. Shaju et al. (2019) investigated the variability of K_d with Chl-*a* in the BOB during the southwest monsoon. They found that the linearity between K_d at wavelengths $\lambda = 412, 443, 490$, and 555 and Chl-*a* was well-described within a linear function, particularly at shorter wavelengths. This highlights the significant influence of phytoplankton in causing variability in $K_d(\lambda)$ in these regions. However, at longer wavelengths, this relationship becomes less prominent due to the strong absorption by seawater.

5.7 Empirical Algorithms for POC

Shaju et al. (2019) evaluated ocean color algorithms for estimating POC in the eastern Arabian Sea (EAS) using in situ POC and inherent optical properties (IOPs). Surface POC correlated significantly with in situ Chl-*a*. Several algorithms, including those based on R_{rs} and $b_{bp}(555)$ (Stramski et al., 2008), $b_{bp}(490)$ (Loisel et al., 2002), and $K_d(490)$ (Gardner et al., 2006), were tested. The R_{rs}-based algorithm (Stramski et al., 2008) yielded the most consistent results though not fully accurate. POC also showed positive correlations with absorption coefficients (a_p, a_{ph}, a_d), suggesting these may serve as optical proxies for POC retrieval via ocean color remote sensing.

5.8 Optical Signatures of Blooms

Dwivedi et al. (2015) investigated the spectral reflectance characteristics of algal blooms in the northeastern AS. They classified phytoplankton blooms into three types: Noctiluca blooms, Diatom blooms, and non-blooms based on the magnitude, slope, and shape of the R_{rs} spectra. For *Noctiluca scintillans* blooms, R_{rs} increases from 440 to 550 nm with a peak at 470 nm and a further increase in reflectance peak at 550 nm (green), which could be attributed to endosymbionts with a green color. In the case of diatoms, the spectral response is relatively flat in this range (440–550 nm), while non-bloom oceanic waters show a decreasing trend of reflectance in this region.

Building on this research, Dias et al. (2020a) investigated the characteristics of dissolved organic matter (DOM) absorption and fluorescence during *Trichodesmium* blooms in the coastal waters off Goa. The blooming of *Trichodesmium,* which belongs to cyanobacteria is a significant event indicating oceanic nitrate deficiency. During peak bloom conditions, the researchers found distinctive absorption peaks in the UV and visible regions. The UV absorption peak at 330 nm was attributed to the presence of UV-absorbing mycosporine-like amino acids (MAAs), while the peak at 360 nm close to the visible region was associated with phycobiliproteins (PBPs) produced by *Trichodesmium*. Furthermore, the spectral fluorescence signatures of *Trichodesmium* exhibited distinct patterns. Traditional peak picking methods revealed three peaks, one protein-like and two humic-like, indicative of different fluorescent components present in the water. Notably, *Trichodesmium* displayed strong protein-like fluorescence with an excitation/emission wavelength of 370/460 nm (C fluorophore, visible humic fluorescence), which is a characteristic signature of cyanobacteria. The fluorescence index (FI = ratio of 470 and 520 nm emissions at 370 nm excitation) values during *Trichodesmium* bloom were notably high (~3), contrasting with the typical range observed in natural waters (1.2–1.8). In degradation experiments simulating *Trichodesmium* bloom degradation, an increase in tryptophan fluorescence was observed (Dias et al., 2020b). This increase correlated with enhanced CDOM absorption and the release of high concentrations of DOM into the coastal waters. These findings underscore the complex interplay between biological processes, optical properties, and environmental dynamics in coastal ecosystems during *Trichodesmium* blooms.

Sathyendranath et al. (1999) investigated phytoplankton photosynthesis rates and specific absorption characteristics during the intermonsoon and monsoon periods. Phytoplankton blooms typically occur after the onset of monsoon winds. Surprisingly, during the Southwest monsoon season, changes in the assimilation number and initial slope of the photosynthesis-light curve decreased, contrary to expectations for explaining blooms in the AS during this time. However, significant differences were observed between intermonsoon and monsoon values of the chlorophyll-specific absorption coefficient (a^*_{ph}), likely due to seasonal changes in cell size and community composition. These changes led to a substantial increase in the maximum quantum yield for photosynthesis during the monsoon compared to the intermonsoon. While seasonal variations in phytoplankton physiological properties were relatively modest, changes in their bio-optical properties, including cell size, were notably strong. This study underscores the importance of adopting an optical perspective in understanding the ocean carbon cycle.

The regional application of the three-component abundance-based model developed by Brewin et al. (2012b) was employed to discern three distinct size classes of phytoplankton in the Northern AS using satellite-derived Chl-*a* data (Sahay et al., 2017). This approach aimed to refine the characterization of phytoplankton size structure in the region. To validate the derived size classes, shipboard data collected during the winter-monsoon seasons of 2009, 2010, and 2011 in the AS were utilized. The effectiveness of the regionally tuned algorithm became evident through the comparison of RMSE values. The average RMSE for the three size components, under the regionally tuned model, notably decreased to 4.7%, compared to the higher value of 9.6% previously reported. This reduction in RMSE signifies an enhanced accuracy in characterizing phytoplankton size classes using the regionally tuned model. Moreover, the regionally tuned model demonstrated its capability to capture monsoonal-driven variations in PSC more effectively than the previous generic model. This improvement highlights the importance of tailoring algorithms to regional conditions for better performance and understanding of ecosystem dynamics.

5.9 CDOM-Bacterial Interaction During Phytoplankton Bloom and Non-Bloom Domains in the NEAS

Basu et al. (2011) investigated CDOM characteristics during Trichodesmium bloom and non-bloom conditions in the northeastern Arabian Sea. Bloom waters showed higher CDOM absorption at 325 nm ($a_{CDOM}(325) = 2.27\ m^{-1}$), lower E_2/E_3 (5.54), and unchanged slope ratio (S_R = 4.09), compared to non-bloom waters

($a_{CDOM}(325) = 0.28\ m^{-1}$, $E_2/E_3 = 20.17$). Bacterial abundance was markedly higher during blooms (94.09×10^8 cells vs. 6.94×10^8). CDOM composition shifted from high-molecular-weight, low-aromatic compounds in bloom regions to low-molecular-weight, high-aromatic DOM in non-bloom areas. Strong Chl-*a* correlations with CDOM ($R^2 = 0.98$, $p < 0.01$) and total bacterial count (TBC) ($R^2 = 0.66$, $p = 0.01$) were observed during blooms. A strong $a_{CDOM}(325)$–TBC correlation ($R^2 = 0.91$, $p < 0.0001$) suggested organic carbon transfer from blooms to heterotrophic bacteria. The study emphasized bacterial processing of bloom-derived CDOM and noted bio-refractory DOM in non-bloom waters. Additionally, mycosporine-like amino acids (λ_{max} 360 nm) and Gram-positive rod-shaped bacteria were detected in bloom samples.

5.10 Seasonal Dynamics of FDOM in the Arabian Sea

Recently, Chari et al. (2025) investigated seasonal variations in the sources of DOM in the EAS using PARAFAC analysis of fluorescence excitation-emission matrix spectra. Spatial differences in protein-like (T and B) and humic-like (C and M) fluorophores during the winter monsoon (WM) and spring intermonsoon (SIM) were found to be linked to physical processes such as convective mixing and the intrusion of BoB waters via the West India Coastal Current (WICC). In the north EAS, DOM derived from phytoplankton during WM shifted toward bacterial-derived DOM during SIM, whereas in the south EAS, the trend was reversed—bacterial DOM dominated during WM due to BoB water intrusion, followed by increased phytoplankton-derived DOM during SIM. The central EAS acted as a transitional zone, where microbial reworking led to the accumulation of DOM, contributing to the development of oxygen minimum zones.

5.11 Optical Dynamics and Variability of Optically Active Substances (OAS) and Their Algorithm Developments in Coastal Regions

Dias et al. (2017) reported a conservative relationship between salinity and $a_{CDOM}(412)$ in summer estuarine waters, suggesting stable CDOM dynamics under typical estuarine conditions. However, in adjacent coastal waters, this relationship deviated due to CDOM losses from photochemical and microbial degradation, reflected in increased $S_{250–600}$ values and decreased spectral slope ratios (S_R), indicating higher molecular weight CDOM and altered optical properties. Later, Dias et al. (2021) observed a strong correlation between $a_{CDOM}(412)$, humic-like fluorescence, and apparent oxygen utilization (AOU) during late southwest monsoon hypoxia, indicating microbial remineralization as a key CDOM source in oxygen-depleted bottom waters. Elevated $a_{CDOM}(412)$ in bottom layers was attributed to benthic fluxes, where organic matter bound to iron oxides was released into the water column, facilitated by sediment–iron interactions.

Menon et al. (2006) investigated the impact of OAS (Chl-*a*, suspended sediments, and CDOM) on in situ R_{rs} in NEAS coastal waters. CDOM significantly influenced optical nonlinearity in the region. They developed spatial and temporal band ratio algorithms using IRS–P4 OCM data to map OAS distribution. Earlier, Menon et al. (2005) found high pre-monsoon variability in $a_{CDOM}(440)$ (1.8–21.6 m^{-1}), with $R_{rs}(\lambda)$ linearly related to absorption and backscattering coefficients. In estuarine and coastal waters, the K_d was lowest, while R_{rs} peaked in the green spectral region. A strong correlation was also observed between $K_d(490)$, K_d PAR, and DCM depth. The $L_u(400)/L_u(670)$ ratio was identified as a proxy for CDOM, highlighting its remote sensing potential. Addressing retrieval limitations in optically complex waters, Shanmugam (2011a) proposed new algorithms to improve conventional band ratio approaches, which often fail due to interference from suspended sediments and dissolved substances in the eastern AS.

5.12 *Optical Characteristics in Upwelled and Oligotrophic Waters*

Coble et al. (1998) observed elevated CDOM concentrations at the DCM in upwelled coastal waters of the northwestern AS, which was attributed to the production of new CDOM associated with the phytoplankton bloom. Biological activity in surface waters of the AS resulted in the production of the putative humic-like fluorophore (M), blue-shifted, and its new associate named as N fluorophore (ex/em, 280/370 nm) and Chl-*a* fluorescence (398/660, designated as P peak). The rapid disappearance of these signals away from the region of the phytoplankton bloom suggested their non-supply. Moreover, oligotrophic waters exhibited extremely low CDOM concentrations, with only UV-absorbing and fluorescing compounds detected. The major sink for CDOM in surface waters was identified as photo-bleaching, a process where CDOM molecules undergo degradation upon exposure to sunlight. This photobleaching phenomenon reduced CDOM concentration and caused a shift in the fluorescence maximum toward shorter wavelengths. Essentially, the fluorescence characteristics of CDOM changed due to this degradation process, with fluorescence peaks moving to shorter wavelengths (Coble et al., 1998).

6 Summary and Conclusions

In the coastal Indian Ocean, particularly in the BOB and the AS, bio-optical studies face significant limitations stemming from environmental factors and methodological challenges. Cloud cover and terrestrial runoff are major hindrances, inducing optical coherence issues in this region. These factors obscure the remote sensing of ocean color, complicating efforts to understand phytoplankton productivity, underwater light fields, and carbon dynamics on synoptic scales within the coastal region. Empirical studies have predominantly focused on coastal and open ocean waters in the AS, leaving the BOB relatively less understudied. The scarcity of research in the BOB is partly attributed to the challenges as mentioned above of cloud cover and terrestrial runoff, which impede consistent data collection and analysis. Additionally, the intricate interplay between various optical properties such as Chl-*a*, CDOM, detritus, and TSM poses complexities that demand sophisticated modeling and analysis techniques which due to the overlapping absorption and scattering characteristics of these constituents, making it challenging to accurately quantify their individual contributions to the observed optical signals. Furthermore, the dynamic nature of coastal environments, influenced by factors like river discharge, tidal currents, and seasonal variability, adds another layer of complexity to bio-optical studies. The photoacclimation conditions of phytoplankton in river-influenced coastal waters of the BOB and AS are not well-understood due to limited observational data and the need for high spatiotemporal resolution measurements. In summary, the bio-optical limitations in the coastal Indian Ocean, particularly in the BOB and AS, stem from a combination of environmental factors, methodological challenges, and the complex interplay of various optical properties. Addressing these limitations requires innovative approaches that integrate remote sensing techniques, in situ observations, and advanced modeling methodologies to improve our understanding of coastal ecosystem dynamics and biogeochemical processes.

In the BOB, anomalies in particle b_{bp} and R_{rs}, induced by cyclones like Hudhud in October 2014, persisted for several months in the coastal regions. These anomalies were attributed to fine suspended particles leaching from coastal deposits formed during the cyclone, hindering the accuracy of Chl-*a* retrieval during this period. The CDOM absorption coefficient $a_{CDOM}(443)$ in the western BOB exhibited an inverse linear relationship with salinity, suggesting a conservative mixing of marine and terrestrial end members (Table 3.3). Higher CDOM levels in the estuary during non-discharge seasons indicated negative estuary conditions due to elevated temperatures and coastal currents transporting CDOM. Seasonal variation in CDOM composition reflected distinct microbial and photodegradation signatures between pre-monsoon and monsoon seasons, bacterial degradation dominating in pre-monsoon seasons and a dominant terrestrial source during monsoon periods. Southwest monsoon influences enrichment of terrestrial-derived organic matter in the northern BOB, which affects optical properties. Empirical algorithms for quantifying phytoplankton size classes and estimating Chl-*a* concentrations in the BOB exhibited varying degrees of accuracy. Superior models can accurately capture nano and microphytoplankton populations but may not capture well the picophytoplankton dynamics. The proposed new regional algorithms as well as new fluorescence-based approaches can tackle the issue of overestimation of Chl-*a* concentrations by global algorithms.

Research in the AS highlights the complex interplay between natural oceanographic processes and anthropogenic factors driving the formation and intensification of *Noctiluca scintillans* (NS) blooms in the northeastern AS (NEAS). While initial debates centered around human-induced pollution as the primary cause of NS blooms, subsequent studies have elucidated the detailed role of natural phenomena such as the WICC in creating favorable conditions for NS proliferation. The comprehensive understanding underscores the need for integrated management strategies that considers both natural variability and anthropogenic impacts on marine ecosystems. The development and refinement of remote sensing algorithms tailored for real-time monitoring of NS blooms represent significant advancements in the field. These algorithms leverage spectral reflectance patterns and SST data to detect and estimate NS abundance, providing valuable tools for timely surveillance and management of NS blooms. By integrating SST into the algorithm, researchers have enhanced its capability to monitor NS blooms across varying environmental conditions, thereby improving the accuracy and reliability of bloom detection.

Various optical classification methods have been employed to characterize water types and optical properties in different marine environments, including estuaries and coastal waters of the Indian Ocean. These classification techniques utilize spectral reflectance patterns and optical parameters to categorize water bodies into distinct types based on their optical signatures. Such classifications offer insights into the spatial and temporal variability of OAS and facilitate the monitoring and understanding of ecosystem dynamics in coastal regions. The evaluation and validation of bio-optical algorithms for estimating parameters such as Chl-*a*, K_d, and POC are crucial for ensuring the accuracy and reliability of remote sensing data. Studies have assessed the performance of various empirical algorithms across different regions and environmental conditions, highlighting the strengths and limitations of each approach. These evaluations contribute to the refinement and optimization of algorithmic models for more accurate remote sensing-based estimations of biogeochemical parameters. Investigating the optical characteristics of phytoplankton blooms, including *Noctiluca scintillans* and *Trichodesmium*, provides valuable insights into their dynamics and ecological impacts. By analyzing spectral reflectance patterns, absorption properties, and fluorescence signatures, researchers can identify distinct optical features associated with different types of blooms. Understanding these optical signatures enhances our ability to detect, monitor, and study phytoplankton blooms, shedding light on their underlying mechanisms and interactions with the surrounding environment.

7 Future Studies and Recommendations

In the realm of optical oceanography research, particularly concerning the Indian Ocean ecosystem, there are several glaring gaps in research and key avenues for future investigation.

- Investigate the coupling of ecological and climate change processes using ocean color models and radiative-transfer models to better understand changes in the

Indian Ocean ecosystem. For example, examine productivity variations in response to monsoon patterns, the Indian Ocean dipole phenomenon, and the heat/absorption budget's influence on bio-optical variability in the Indian Ocean north of 8°S.

- Explore the impact of global warming-induced stratification on Indian Ocean productivity.
- Compare the global anthropogenic carbon dioxide sink with biological productivity and assess ocean acidification trends in the Indian Ocean.
- Evaluate biological carbon pathways by deciphering bio-optical signatures. Utilize Argo particulate b_{bp} and Chl-*a* profiles to quantify the export of Particulate Organic Carbon (POC) in the Indian Ocean region.
- Develop PP-SST-Light bio-optical models to predict climate change effects on carbon dynamics using long-term bio-optical data.
- Exploring the relationship between vertical profiles of the Deep Chlorophyll Maximum (DCM) and b_{bp} profiles of Chl-*a* in the upper ocean layer can provide valuable insights into the physiological state of phytoplankton.

Acknowledgements The authors thank the NCPOR and INCOIS directors for their encouragement. INCOIS supported this work through grant No. F&A: XII: D2:023 and CSIR-HRDG for financial support as CSIR-Research Associate (RA) [funding no: 09/002 (510)/2016 EMR-I]. This is the NCPOR contribution number B-3/2025-26.

References

Aguilar-Maldonado, J. A., Santamaría-del-Angel, E., Gonzalez-Silvera, A. G., Cervantes Rosas, O. D., Lopez, L. M., Gutierrez-Magness, A., & Sebasti a-Frasquet, M. T. (2017). Identification of phytoplankton blooms under the index of inherent optical properties (IOPindex). *Proceedings, 2*, 187. https://doi.org/10.3390/ecws-2-04956

Albin, K. J., Jyothibabu, R., Alok, K. T., Santhikrishnan, S., Sarath, S., Sudheesh, V., Sherin, C. K., Balachandran, K. K., AshaDevi, C. R., & Gupta, G. V. M. (2022). Distinctive phytoplankton size responses to the nutrient enrichment of coastal upwelling and winter convection in the eastern Arabian Sea. *Progress in Oceanography, 203*, 102779. https://doi.org/10.1016/j.pocean.2022.102779

Ashraf, P. M., Shaju, S. S., Gayatri, D., et al. (2013). In situ time series estimation of Downwelling diffuse attenuation coefficient at Southern Bay of Bengal. *Journal of the Indian Society of Remote Sensing, 41*, 725–730. https://doi.org/10.1007/s12524-012-0244-1

Ashraf, P. M., Shaju, S. S., Gayatri, D., Minu, P., & Meenakumari, B. (2013). In Situ Time Series Estimation of Downwelling Diffuse Attenuation Coefficient at Southern Bay of Bengal. *Journal of Indian Society and Remote Sensing*, 41, 725–730. https://doi.org/10.1007/s12524-012-0244-1

Babin, M., Stramski, D., Ferrari, G. M., Claustre, H., Bricaud, A., Obolensky, G., & Hoepffner, N. (2003). Variations in the light absorption coefficients of phytoplankton, nonalgal particles, and dissolved organic matter in coastal waters around. Europe. *Journal of Geophysical Research: Oceans, 108*. https://doi.org/10.1029/2001jc000882

Balch, W. M., Drapeau, D. T., Fritz, J. J., Bowler, B. C., & Nolan, J. (2001). Optical backscattering in the Arabian Sea - Continuous underway measurements of particulate inorganic and

organic carbon. *Deep. Research Part I Oceanography Research*, 48, 2423–2452. https://doi.org/10.1016/S0967-0637(01)00025-5
Barocio-Leon, O. A., Millan-Nunez, R., Santamaría-del-Angel, E., Gonzalez-Silvera, A., & Trees, C. C. (2006). Spatial variability of phytoplankton absorption coefficients and pigments off Baja California during November 2002. *Journal of Oceanography, 62*, 873–885. https://doi.org/10.1007/s10872-006-0105-z
Basu, S., Matondkar, S., & Furtado, I. (2011). Enumeration of bacteria from a Trichodesmium spp. bloom of the eastern Arabian Sea: Elucidation of their possible role in biogeochemistry. *Journal of Applied Phycology, 23*, 309–319.
Behrenfeld, M. J., O'Malley, R. T., Siegel, D. A., McClain, C. R., Sarmiento, J. L., Feldman, G. C., Milligan, A. J., Falkowski, P. G., Letelier, R. M., & Boss, E. S. (2006). Climate-driven trends in contemporary ocean productivity. *Nature*, 444, 752–755. https://doi.org/10.1038/nature05317
Bianchi, D., Dunne, J. P., Sarmiento, J. L., & Galbraith, E. D. (2012). Data-based estimates of sub-oxia, denitrification, and N2O production in the ocean and their sensitivities to dissolved O2. *Deep Sea Research Part I*, 51, 1159–1168. https://doi.org/10.1016/j.dsr.2004.03.009
Bonelli, A. G., Vantrepotte, V., Jorge, D. S. F., et al. (2021). Colored dissolved organic matter absorption at global scale from ocean color radiometry observation: Spatio-temporal variability and contribution to the absorption budget. *Remote Sensing of Environment,* 265:. https://doi.org/10.1016/j.rse.2021.112637
Bopp, L., Resplandy, L., Orr, J. C., Doney, S. C., Dunne, J. P., Gehlen, M., Halloran, P., Heinze, C., Ilyina, T., Séférian, R., Tjiputra, J., & Vichi, M. (2013). Multiple stressors of ocean ecosystems in the 21st century: projections with CMIP5 models, *Biogeosciences,* 10, 6225–6245, https://doi.org/10.5194/bg-10-6225-2013, 2013.
Bowers, D. G., Md-Suffian, I., & Mitchelson-Jacob, E. G. (2012). Bio-optical properties of East Coast Malaysia waters in relation to remote sensing of chlorophyll. *International Journal of Remotesensing*, 33, 150–169. https://doi.org/10.1080/01431161.2011.584917
Bracher, A., Vountas, M., Dinter, T., & Burrows, J. P. (2009). Quantitative observation of cyanobacteria and diatoms from space using PhytoDOAS on SCIAMACHY data. *Biogeosciences*, 6, 751–764. https://doi.org/10.5194/bg-6-751-2009
Breves, W., & Reuter, R. (2000). Bio-optical properties of gelbstoff in the Arabian Sea at the onset of the southwest monsoon. *Proceedings of the Indian Academy of Sciences, Earth and Planetary Sciences*, 109, 415–425. https://doi.org/10.1007/BF02708329
Brewin, R. J. W., Sathyendranath, S., Hirata, T., Lavender, S. J., Barciela, R. M., & Hardman Mountford, N. J. (2010). A three-component model of phytoplankton size class for the Atlantic Ocean. *Ecological Modelling, 221*, 1472–1483. https://doi.org/10.1016/J.ECOLMODEL.2010.02.014
Brewin, R. J. W., DallOlmo, G., Sathyendranath, S., & Hardman-Mountford, N. J. (2012a). Particle backscattering as a function of chlorophyll and phytoplankton size structure in the open-ocean. *Optics Express, 20*, 17632–17652. https://doi.org/10.1364/OE.20.017632
Brewin, R. J. W., Hirata, T., Hardman-Mountford, N. J., Lavender, S. J., Sathyendranath, S., & Barlow, R. (2012b). The influence of the Indian Ocean dipole on interannual variations in phytoplankton size structure as revealed by earth observation. *Deep-Sea Research Part II, 77-80*, 117–127. https://doi.org/10.1016/j.dsr2.2012.04.009
Brewin, R. J. W., Sathyendranath, S., Jackson, T., Barlow, R., Brotas, V., Airs, R., & Lamont, T. (2015). Influence of light in the mixed layer on the parameters of a three-component model of phytoplankton size structure. *Remote Sensing of Environment, 168*, 437–450.
Bricaud, A., & Stramski, D. (1990). Spectral absorption coefficients of living phytoplankton and nonalgal biogenous matter: A comparison between the Peru upwelling area and the Sargasso Sea. *Limnology and Oceanography, 35*, 562–582. https://doi.org/10.4319/lo.1990.35.3.0562
Bricaud, A., Babin, M., Morel, A., & Claustre, H. (1995). Variability in the chlorophyll specific absorption coefficients of natural phytoplankton: Analysis and parameterization. *Journal of Geophysical Research, 100*, 13321–13332. https://doi.org/10.1029/95JC00463

Bricaud, A., Claustre, H., Ras, J., & Oubelkheir, K. (2004). Natural variability of phytoplanktonic absorption in oceanic waters: Influence of the size structure of algal populations. *Journal of Geophysical Research. Oceans, 109*, C11010. https://doi.org/10.1029/2004JC002419

Carritt, D. E., & Carpenter, J. H. (1966). Comparison and evaluation of currently employed modification of Winkler method for determination of dissolved oxygen in seawater. A NASCO report. *Journal of Marketing Research, 24*, 287–318.

Chari, N. V. H. K., & Pandi, S. R. (2017). Characterization of humic substances and their distribution of XAD fractions by absorption spectroscopy in the Godavari estuary, India. *Current Science, 113*(2), 299–303.

Chari, N. V. H. K., Sarma, N. S., Pandi, S. R., & Murthy, K. N. (2012). Seasonal and spatial constraints of fluorophores in the midwestern bay of Bengal by PARAFAC analysis of excitation-emission matrix spectra. *Estuarine, Coastal and Shelf Science, 100*, 162–171.

Chari, N. V. H. K., Suman, K., Sarma, N. S., Pandi, S. R., Chiranjeevulu, G., Kiran, R., & Koduru, U. (2013). Fluorescence and absorption characteristics of dissolved organic matter excreted by phytoplankton species of western bay of Bengal under axenic laboratory conditions. *Journal of Experimental Marine Biology and Ecology, 445*, 148–155.

Chari, N. V. H. K., Sarma, N. S., Pandi, S. R., Chiranjeevulu, G., Kiran, R., Murthy, K. N., & Venkatesh, P. (2016). Fluorescent dissolved organic matter dynamics in the coastal waters off the central east Indian coast (Bay of Bengal). *Environment and Ecology Research, 4*, 13–20.

Chari, N. V. H. K., Pandi, S. R., Kanuri, V. V., & Basuri, C. K. (2019). Structural variation of coloured dissolved organic matter during summer and winter seasons in a tropical estuary: A case study. *Marine Pollution Bulletin, 149*, 110563. https://doi.org/10.1016/j.marpolbul.2019.110563

Chari, N. V. H. K., Gupta, G. V. M., Sherin, C. K., Ramu, C. V., Reddy, B. B., & Kumar, B. S. K. (2025). Variability of dissolved organic matter sources in the eastern Arabian Sea: Impacts of physical and biological processes. *Ocean Dynamics, 75*(3). https://doi.org/10.1007/s10236-024-01648-1

Chauhan, P., & Raman, M. (2017). Satellite remote sensing for ocean biology: An Indian perspective. *Proceedings of the National Academy of Sciences, India Section A: Physical Sciences, 87*, 629–640.

Chauhan, P., Mohan, M., Nayak, S. R., & Navalgund, R. R. (2002a). Comparison of chlorophyll-*a* algorithms using bio-optical data collected in the Arabian Sea. *Journal of the Indian Society of Remote Sensing, 30*, 87–94.

Chauhan, P., Mohan, M., Sarangi, R. K., Kumari, B., Nayak, S., & Matondkar, S. G. P. (2002b). Surface chlorophyll estimation in the Arabian Sea using IRS-P4 OCM Ocean color monitor (OCM) satellite data. *International Journal of Remote Sensing, 23*, 1663–1676. https://doi.org/10.1080/01431160110075866

Chauhan, P., Sahay, A., Rajawat, A. S., & Nayak, S. (2003). Remote sensing of diffuse attenuation coefficient (K490) using IRS-P4 ocean colour monitor (OCM) sensor. *Indian Journal of Geo-Marine Sciences, 32*, 279–284.

Chen, Z., Hu, C., Conmy, R. N., MullerKarger, F., & Swarzenski, P. (2007). Colored dissolved organic matter in Tampa Bay, Florida. *Marine Chemistry, 104*, 98–109. https://doi.org/10.1016/j.marchem.2006.12.007

Chen, Z., Li, Y., Pan, J. (2004). Distributions of colored dissolved organic matter and dissolved organic carbon in the Pearl River Estuary, China. Continental Shelf Research, 24, 1845–1856. https://doi.org/10.1016/j.csr.2004.06.011

Chiranjeevulu, G., Murty, K. N., Sarma, N. S., Kiran, R., Chari, N. V. H. K., Pandi, S. R., Venkatesh, P., Annapurna, C., & Nageswara Rao, K. (2014). Coloured dissolved organic matter signature and phytoplankton response in a coastal ecosystem during mesoscale cyclonic (cold core) eddy. *Marine Environmental Research, 98*, 49–59.

Ciotti, A. M., Cullen, J. J., & Lewis, M. R. (2002). Assessment of the relationships between dominant cell size in natural phytoplankton communities and the spectral shape of absorption coefficient. *Limnology and Oceanography, 47*, 404–417.

Clementson, L. A., Parslow, J. S., Turnbull, A. R., McKenzie, D. C., & Rathbone, C. E. (2001). Optical properties of waters in the Australasian sector of the we studied in situ parameters layer chl a concentrations and showed a general decrease by the phytoplankton were observed and south exceeded pigment the sea-viewing generally underestimated. *Journal of Geophysical Research, 106*, 31611–31625.

Coble, P. G. (2007). Marine optical biogeochemistry: The chemistry of ocean color. *Chemical Reviews, 107*, 402–418.

Coble, P. G., Del Castillo, C. E., & Avril, B. (1998). Distribution and optical properties of CDOM in the Arabian Sea during the 1995 southwest monsoon. *Deep-Sea Research Part II, 45*, 2195–2223.

D'Sa, E. J., & Miller, R. L. (2005). Bio-optical properties of coastal waters. In R. L. Miller, C. E. DelCastillo, & B. A. Mckee (Eds.), *Remote sensing of coastal aquatic environments* (pp. 129–155). Springer.

Das, S., & Hazra, S. (2019). Vertical distribution of optically active substances (OAS) in the coastal sea off Indian Sundarban during post monsoon. *Marine Biology Research*. https://doi.org/10.1080/17451000.2019.1607387

Das, S., Das, I., Giri, S., et al. (2017a). Chromophoric dissolved organic matter (CDOM) variability over the continental shelf of the northern Bay of Bengal. *Oceanologia, 59*, 271–282. https://doi.org/10.1016/j.oceano.2017.03.002

Das, S., Hazra, S., Giri, S., et al. (2017b). Light absorption characteristics of chromophoric dissolved organic matter (CDOM) in the coastal waters of northern Bay of Bengal during winter season. *Indian Journal of Geo-Marine Sciences, 46*, 884–892.

Das, S., Hazra, S., Lotlikar, A. A., Das, I., Giri, S., Chanda, A., Akhand, A., Maity, S., & Kumar, T. S. (2016). Delineating the relationship between chromophoric dissolved organic matter (CDOM) variability and biogeochemical parameters in a shallow continental shelf. *Egyptian Journal of Aquatic Research*, 42, 241–248. https://doi.org/10.1016/j.ejar.2016.08.001

Del Castillo, C. E., & Coble, P. G. (2000). Seasonal variability of the colored dissolved organic matter during the 1994–1995 NE and SW monsoons in the Arabian Sea. *Deep-Sea Research Part II, 47*, 1563–1579. https://doi.org/10.1016/S0967-0645(99)00154-X

Del Castillo, C. E., Coble, P. G., Morell, J. M., López, J. M., & Corredor, J. E. (1999). Analysis of the optical properties of the Orinoco River plume by absorption and fluorescence spectroscopy. *Marine Chemistry*, 66, 35–51. https://doi.org/10.1016/S0304-4203(99)00023-7

Deppeler, S. L., & Davidson, A. T. (2017). Southern Ocean phytoplankton in a changing climate. *Frontiers in Marine Science, 4*. https://doi.org/10.3389/fmars.2017.00040

Devred, E. S., Sathyendranath, V., Stuart, H., Ulloa, M. O., & Platt, T. (2006). A two-component model of phytoplankton absorption in the open ocean: Theory and applications. *Journal of Geophysical Research, 111*, C03011. https://doi.org/10.1029/2005JC002880

Dias, A., & Kurian, S. (2022). Characterization of colored dissolved organic matter along the western continental shelf of India during the seasonal hypoxia. *Estuarine, Coastal and Shelf Science, 265*, 107714.

Dias, A., Suresh, T., Sahay, A., & Chauhan, P. (2017). Contrasting characteristics of colored dissolved organic matter of the coastal and estuarine waters of Goa during summer. *Indian Journal of Geo-Marine Sciences, 46*, 860–870.

Dias, A., Kurian, S., & Thayapurath, S. (2020a). Influence of environmental parameters on bio-optical characteristics of colored dissolved organic matter in a complex tropical coastal and estuarine region. *Estuarine, Coastal and Shelf Science, 242*, 106864.

Dias, A., Kurian, S., & Thayapurath, S. (2020b). Optical characteristics of colored dissolved organic matter during blooms of *Trichodesmium* in the coastal waters off Goa. *Environmental Monitoring and Assessment, 192*, 1–18.

Dias, A., Kurian, S., Thayapurath, S., & Pratihary, A. K. (2021). Variations of colored dissolved organic matter in the Mandovi estuary, Goa, during spring inter-monsoon: A comparison with COVID-19 outbreak imposed lockdown period. *Frontiers in Marine Science, 8*, 638583. https://doi.org/10.3389/fmars.2021.638583

Dwivedi, R., Rafeeq, M., Smitha, B. R., Padmakumar, K. B., Thomas, L. C., Sanjeevan, V. N., Prakash, P., & Raman, M. (2015). Species identification of mixed algal bloom in the northern Arabian Sea using remote sensing techniques. *Environmental Monitoring and Assessment, 187*(2), 51. https://doi.org/10.1007/s10661-015-4291-2. Epub 2015 Feb 1.

Falkowski, P. G. (1994). The role of phytoplankton photosynthesis in global biogeochemical cycles. *Photosynthesis Research, 39*, 235–258. https://doi.org/10.1007/BF00014586

Ferreira, A., Ciotti, A. M., Mendes, C. R. B., Uitz, J., & Bricaud, A. (2017). Phytoplankton light absorption and the package effect in relation to photosynthetic and photoprotective pigments in the northern tip of Antarctic peninsula. *Journal of Geophysical Research: Oceans, 122*, 7344–7363. https://doi.org/10.1002/2017JC012964

Follows, M. J., Dutkiewicz, S., Grant, S., & Chisholm, S. W. (2007). Emergent biogeography of microbial communities in a model ocean. *Science*, 315, 1843–1846. https://doi.org/10.1126/science.1138544

Fujiki, T., & Taguchi, S. (2002). Variability in chlorophyll a specific absorption coefficient in marine phytoplankton as a 566 function of cell size and irradiance. *Journal of Plankton Research, 24*, 859–874. https://doi.org/10.1093/plankt/24.9.859

Gardner, W. D., Mishonov, A. V., & Richardson, M. J. (2006). Global POC concentrations from in-situ and satellite data. *Deep-Sea Research Part II, 53*, 718–740.

Gomes, H. do R., S. deRada, J. I. Goes, & F. Chai. (2016). Examining features of enhanced phytoplankton biomass in the Bay of Bengal using a coupled physical-biological model. *Journal of Geophysical Research: Oceans*, 121, 5112–5133, https://doi.org/10.1002/2015JC011508

Gomes, H., Goes, J., Matondkar, S., et al. (2014). Massive outbreaks of Noctiluca scintillans blooms in the Arabian Sea due to spread of hypoxia. *Nature Communications, 5*, 4862. https://doi.org/10.1038/ncomms5862

Gordon, H. R., & Morel, A. (1983). *Remote assessment of ocean color for interpretation of satellite visible imagery: A review* (p. 44). Springer.

Gordon, H. R., Brown, O. B., & Jacobs, M. M. (1975). Computed relationships between the inherent and apparent optical properties of a flat homogeneous ocean. *Applied Optics, 14*, 417–427.

Grasshof, K., Ehrhardt, M., & Kremling, K. (1999). *Methods of seawater analysis* (3rd ed., p. 6). Verlag Chemie.

Green, S. A., Blough, N. V. (1994). Optical absorption and fluorescence properties of chromophoric dissolved organic matter in natural waters. *Limnology and Oceanography,* 39, 1903–1916. https://doi.org/10.4319/lo.1994.39.8.1903

Han, B., Markku, V., Koponen, J., & Milan, S. (2000). Effect of photoinhibition on algal photosynthesis: A dynamic model. *Journal of Plankton Research, 22*, 865–885.

Hansen, A. M., Kraus, T. E. C., Pellerin, B. A., Fleck, J. A., Downing, B. D., & Bergamaschi, B. A. (2016). Optical properties of dissolved organic matter (DOM): Effects of biological and photolytic degradation. *Limnology and Oceanography*, 61, 1015–1032. https://doi.org/10.1002/lno.10270

Helly, J. J., Levin, L. A. (2004). Global distribution of naturally occurring marine hypoxia on continental margins. *Deep Sea Research Part I,* 51, 1159–1168. https://doi.org/10.1016/j.dsr.2004.03.009

Helms, J. R., Stubbins, A., Ritchie, J. D., et al. (2008). Absorption spectral slopes and slope ratios as indicators of molecular weight, source, and photobleaching of chromophoric dissolved organic matter. *Limnology and Oceanography, 53*, 955–969. https://doi.org/10.4319/lo.2008.53.3.0955

Hirata, T., Aiken, J., Hardman-Mountford, N., Smyth, T. J., & Barlow, R. G. (2008). An absorption model to determine phytoplankton size classes from satellite ocean colour. *Remote Sensing of Environment, 112*, 3153–3159. https://doi.org/10.1016/j.rse.2008.03.011

Hood, R.R., Laws, E.A., Armstrong, R.A., Bates, N.R., Brown, C.W., Carlson, C.A. et al. (2006). Pelagic functional group modeling: Progress, challenges and prospects. *Deep-Sea Research II,* 53, 459–512.

Hong, H., Wu, J., Shang, S., & Hu, C. (2005). Absorption and fluorescence of chromophoric dissolved organic matter in the Pearl River Estuary, South China. *Marine Chemistry*, 97, 78–89. https://doi.org/10.1016/j.marchem.2005.01.008

Hur, J., & Cho, J. (2012). Prediction of BOD, COD, and total nitrogen concentrations in a typical urban river using a fluorescence excitation-emission matrix with PARAFAC and UV absorption indices. *Sensors*, 972–986. https://doi.org/10.3390/s120100972

IOCCG. (2000). Remote sensing of ocean colour in coastal, and other optically complex, waters. In S. Sathyendranath (Ed.), *Reports of the International Ocean-colour co-ordinating group* (Vol. 3).

IOCCG. (2014). Phytoplankton functional types from space. Sathyendranath, S. (ed.), reports of the international ocean-colour coordinating group, No. 15, *IOCCG*, Dartmouth, Canada.

IOCCG. (2006). Remote sensing of inherent optical properties: Fundamentals, tests of algorithms, and applications. In Z. P. Lee (Ed.), *Reports of the International Ocean-Colour Coordinating Group, No. 5* (p. 126). International Ocean-Colour Coordinating Group (IOCCG). https://doi.org/10.25607/OBP-96

Jayaram, C., Udaya Bhaskar, T. V. S., Chacko, N., Satya Prakash, & Rao, K. H. (2021). Spatio-temporal variability of chlorophyll in the northern Indian Ocean: A biogeochemical argo data perspective. *Deep-Sea Research Part II, 183*, 104928.

Jerlov, N. G. (1976). *Marine optics* (Vol. 14). Elsevier.

Jyothibabu, R., Vinayachandran, P. N., Madhu, N. V., Robin, R. S., Karnan, C., Jagadeesan, L., & Anjusha, A., (2015). Phytoplankton size structure in the southern Bay of Bengal modified by the Summer Monsoon Current and associated eddies: Implications on the vertical biogenic flux. *Journal of Marine Systems*, 143, 98–119. https://doi.org/10.1016/j.jmarsys.2014.10.018

Keith, D. J., Yoder, J. A., & Freeman, S. A. (2002). Spatial and temporal distribution of coloured dissolved organic matter (CDOM) in Narragansett Bay, Rhode Island: Implications for phytoplankton in coastal waters. *Estuarine and Coastal Shelf Sciences,* 55, 705–717. https://doi.org/10.1006/ecss.2001.0922

Kirk, J. T. O. (2011). *Light and photosynthesis in aquatic ecosystems* (3rd ed.). Cambridge University Press.

Kishino, M., Takahashi, M., Okami, N., & Ichimura, S. (1985). Estimation of the spectral absorption coefficients of phytoplankton in the sea. *Bulletin of Marine Science*, 37, 634–642.

Kowalczuk, P., Stoń-Egiert, J., Cooper, W. J., et al. (2005). Characterization of chromophoric dissolved organic matter (CDOM) in the Baltic Sea by excitation emission matrix fluorescence spectroscopy. *Marine Chemistry, 96*, 273–292. https://doi.org/10.1016/j.marchem.2005.03.002

Krishna, A. V., Tirkey, A., Raman, M., et al. (2024). Exploring the interplay between phytoplankton taxonomy, pigments, and optical properties in Ratnagiri coastal waters using a multi-parameter approach. *Discover Oceans, 1*, 23. https://doi.org/10.1007/s44289-024-00025-6

Krumhardt, K. M., Lovenduski, N. S., Iglesias-Rodriguez, M. D., & Kleypas, J. A. (2017). Coccolithophore growth and calcification in a changing ocean. *Progress in Oceanography,* 159, 276–295. https://doi.org/10.1016/j.pocean.2017.10.007

Lakowicz, J. R. (2006). *Principles of fluorescence spectroscopy* (3rd ed.). Springer. https://doi.org/10.1007/978-0-387-46312-4

Le Quéré, C., Harrison, S. P., Prentice, C. I., Buitenhuis, E. T., Aumonts, O., Bopp, L., et al. (2005). Ecosystem dynamics based on plankton functional types for global ocean biogeochemistry models. *Global Change Biology*, 11, 2016–2040.

Lee, Z., Carder, K. L., & Arnone, R. A. (2002). Deriving inherent optical properties from water color: A multiband quasi-analytical algorithm for optically deep waters. *Applied Optics, 41*, 5755. https://doi.org/10.1364/ao.41.005755

Lee, Z. P., Darecki, M., Carder, K. L., Davis, C. O., Stramski, D., & Rhea, W. J. (2005). Diffuse attenuation coefficient of downwelling irradiance: An evaluation of remote sensing methods. *Journal of Geophysical Research: Oceans, 110*, 1–9. https://doi.org/10.1029/2004JC002573

Lévy M, Resplandy L, Palter JB, Couespel D, & Lachkar Z. (2022). The crucial contribution of mixing to present and future ocean oxygen distribution. In Ocean Mixing: Drivers, Mechanisms and Impacts, ed. M Meredith, AN Garabato, pp. 329–44.

Lohrenz, S. E., Weidemann, A. D., & Tuel, M. (2003). Phytoplankton spectral absorption as influenced by community size structure and pigment composition. *Journal of Plankton Research, 25*, 35–61.

Loisel, H., Meriaux, X., Berthon, J., & Poteau, A. (2007). Investigation of the optical backscattering to scattering ratio of marine particles in relation to their biogeochemical composition in the eastern english channel and southern north sea. *Limnology and Oceanography,* 52 (2), 739–752. JRC31896.

Loisel, H., Nicolas, J. M., Deschamps, P. Y., & Frouin, R. (2002). Seasonal and inter-annual variability of particulate organic matter in the global ocean. *Geophysical Research Letters, 29*, 2196. https://doi.org/10.1029/2002GL015948

Lotliker, A. A., Baliarsingh, S. K., Sahu, K. C., & Kumar, T. S. (2020). Long-term chlorophyll-a dynamics in tropical coastal waters of the western bay of Bengal. *Environmental Science and Pollution Research, 27*(6), 6411–6419. https://doi.org/10.1007/s11356-019-07403-0

Lotliker, A. A., Baliarsingh, S. K, Trainer, V. L, et al (2018). Characterization of oceanic Noctiluca blooms not associated with hypoxia in the Northeastern Arabian Sea. *Harmful Algae*, 74, 46–57. https://doi.org/10.1016/j.hal.2018.03.008

Lozano, I., & Alados, I. (2023). Improving the Estimation of the Diffuse Component of Photosynthetically Active Radiation (PAR). *Journal of Geophysical Research : Atmospheres*, 1–16. https://doi.org/10.1029/2023JD039256

Matsuoka, A., Bricaud, A., Benner, R., Para, J., Sempéré, R., Prieur, L., Bélanger, S., & Babin, M. (2011). Tracing the transport of colored dissolved organic matter in water masses of the Southern Beaufort Sea: relationship with hydrographic characteristics. *Biogeosciences Discussion*, 8, 11003–11040. https://doi.org/10.5194/bgd-8-11003-2011

Menon, H. B., Lotliker, A. A., & Nayak, S. R. (2005). Pre-monsoon bio-optical properties in estuarine, coastal and Lakshadweep waters. *Estuarine, Coastal and Shelf Science, 63*, 211–223.

Menon, H. B., Lotliker, A. A., Moorthy, K. K., & Nayak, S. R. (2006). Variability of remote sensing reflectance and implications for optical remote sensing—A study along the eastern and northeastern waters of Arabian Sea. *Geophysical Research Letters, 33*, L15602. https://doi.org/10.1029/2006GL026026

Minu, P., Lotliker, A. A., Shaju, S. S., Santhosh Kumar, B., Ashraf, P. M., & Meenakumari, B. (2014). Effect of optically active substances and atmospheric correction schemes on remote sensing reflectance at a coastal site off Kochi. *International Journal of Remote Sensing, 35*, 5434–5447. https://doi.org/10.1080/01431161.2014.926420

Minu, P., Lotliker, A. A., Shaju, S. S., Ashraf, P. M., Kumar, T. S., & Meenakumari, B. (2016). Performance of operational satellite bio-optical algorithms in different water types in the southeastern Arabian Sea. *Oceanologia, 58*, 317–326. https://doi.org/10.1016/j.oceano.2016.05.005

Miranda, J., Lotliker, A. A., Sanjiba Kumar, B., Jena, A. K., Alakes, S., Sahu, K. C., & Srinivasa Kumar, T. (2021). Satellite estimates of the long-term trend in phytoplankton size classes in the coastal waters of north-western Bay of Bengal. *Oceanologia, 63*, 40–50.

Mobley, C. D. (2022). *The oceanic optics book* (p. 924). International Ocean Colour Coordinating Group (IOCCG). https://doi.org/10.25607/OBP-1710

Mobley, C. D., Stramski, D., Bissett, W. P., & Boss, E. (2004). Optical modeling of ocean waters: Is the case 1—case 2 classification still useful? *Oceanography, 17*, 60–67.

Monolisha, S., Platt, T., Sathyendranath, S., Jayasankar, J., George, G., & Jackson, T. (2018). Optical classification of the coastal waters of the northern Indian Ocean. *Frontiers in Marine Science, 5*, 87. https://doi.org/10.3389/fmars.2018.00087

Moore, T. S., Campbell, J. W., & Feng, H. (2001). A fuzzy logic classification scheme for selecting and blending Satellite Ocean color algorithms. *IEEE Transactions on Geoscience and Remote Sensing, 39*, 8. https://doi.org/10.1109/36.942555

Moore, T. S., Campbell, J. W., & Dowell, M. D. (2009). A class-based approach to characterizing and mapping the uncertainty of the MODIS Ocean chlorophyll product. *Remote Sensing of Environment, 113*, 2424–2430.

Morel, A. (1988). Optical modeling of the upper ocean in relation to itsbiogenous matter content (case I waters). *Journal of Geophysical Research, 93*, 10749–10768.

Morel, A., & Maritorena, S. (2001). Bio-optical properties of oceanic waters: A reappraisal. *Journal of Geophysical Research, 106*(C4), 7163–7180. https://doi.org/10.1029/2000JC000319

Morel, A., & Prieur, L. (1977). Analysis of variations in ocean color. *Limnology and Oceanography, 22*, 709–722. https://doi.org/10.4319/lo.1977.22.4.0709

Murphy, K. R., Butler, K. D., Spencer, R. G. M., et al. (2010). Measurement of dissolved organic matter fluorescence in aquatic environments: An interlaboratory comparison. *Environmental Science & Technology, 44*, 9405–9412. 618. https://doi.org/10.1021/es102362t

Murphy, K. R., Stedmon, C. A., Waite, T. D., & Ruiz, G. M. (2008). Distinguishing between terrestrial and autochthonous organic matter sources in marine environments using fluorescence spectroscopy. *Marine Chemistry*, 108, 40–58. https://doi.org/10.1016/j.marchem.2007.10.003

Nagamani, P. V., Chauhan, P., & Dwivedi, R. M. (2008). Development of chlorophyll-*a* algorithm for ocean colour monitor onboard OCEANSAT-2 satellite. *IEEE Geoscience and Remote Sensing Letters, 5*, 3.

Nagamani, P., Hussain, M., Choudhury, S., Panda, C., Sanghamitra, P., Kar, R., Das, A., Ramana, I., & Rao, K. (2013). Validation of chlorophyll-a algorithms in the coastal waters of bay of Bengal initial validation results fromOCM-2. *Journal of the Indian Society of Remote Sensing, 41*, 117–125.

Nagamani, P. V., Preethi Latha, Rao, K. H., Suresh, T., Choudhury, S. B., Dutt, C. B. S., et al. (2015). Setting of cloud albedo in the atmospheric correction procedure to generate the ocean colour data products from OCM-2. *Journal of the Indian Society of Remote Sensing, 43*(2), 439–444.

Nagamani, P. V., Preethi Latha, T., Bhavani, I., Rao, V. G., Raman, Y. U., Amminedu, M., et al. (2017). Optical detection of daitom bloom in the coastal waters of bay of Bengal using Oceansat-2 OCM. *National Academy of Sciences, 1*(2). https://doi.org/10.1007/s40010-017-0455-5

Neukermans, G., Bach, L. T., Butterley, A., Sun, Q., Claustre, H., & Fournier, G. R. (2023). Quantitative and mechanistic understanding of the open ocean carbonate pump: Perspectives for remote sensing and autonomous in situ observation. *Earth-Science Reviews, 239*. https://doi.org/10.1016/j.earscirev.2023.104359

Nieke, B., Reuter, R., Heuermann, R., Wang, H., Babin, M., & Therriault, J. C. (1997). Light absorption and fluorescence properties of chromophoric dissolved organic matter (CDOM), in the St. Lawrence Estuary (Case 2 waters). *Continental Shelf Research*, 17, 235–252. https://doi.org/10.1016/S0278-4343(96)00034-9

Niloy, N. M., Habib, S. M. A., Islam, M. I., Haque, M., Shammi, M., & Tareq, S. M. (2023). Distribution, characteristics and fate of fluorescent dissolved organic matter (FDOM) in the Bay of Bengal. *Marine Pollution Bulletin, 195*, 115467. https://doi.org/10.1016/j.marpolbul.2023.115467

O'Reilly, J. E., & Werdell, P. J. (2019). Chlorophyll algorithms for ocean color sensors—OC4, OC5 & OC6. *Remote Sensing of Environment, 229*, 32–47. https://doi.org/10.1016/j.rse.2019.04.021

O'Reilly, J. E., Maritorena, S., Mitchell, B. G., Siegel, D. A., Carder, K. L., Garver, S. A., Kahru, M., & McClain, C. (1998). Ocean color chlorophyll algorithms for SeaWiFS. *Journal of Geophysical Research, Oceans, 103*(C11), 24937–24953. https://doi.org/10.1029/98JC02160

Oubelkheir, K., Claustre, H., Sciandra, A., & Babin, M. (2005). Bio-optical and biogeochemical properties of different trophic regimes in oceanic waters. *Limnology and Oceanography, 50*(6), 1795–1809. https://doi.org/10.4319/lo.2005.50.6.1795

Pandi, S. R., Kiran, R., Sarma, N. S., Srikanth, A. S., Sarma, V. V. S. S., Krishna, M. S., Bandyopadhyay, D., Prasad, V. R., Acharyya, T., & Reddy, K. G. (2014). Contrasting phytoplankton community structure and associated light absorption characteristics of the western bay of Bengal. *Ocean Dynamics, 64*, 89–101.

Pandi, S. R., Chiranjeevulu, G., Kiran, R., Chari, N. V. H. K., Muralikrishna, R., Srinivasarao, B., Lotliker, A. A., & Sarma, N. S. (2017). Contrasting bio-optical characteristics of coastal water prior to and in the aftermath of a tropical super cyclone. *International Journal of Remote Sensing, 38*, 3519–3530.

Pandi, S. R., Baliarsingh, S. K., Lotliker, A. A., Sarma, N. S., & Tripathy, S. C. (2020). Empirical relationships for remote sensing reflectance and Noctiluca scintillans cell density in the northeastern Arabian Sea. *Marine Pollution Bulletin, 161*, 111770. https://doi.org/10.1016/j.marpolbul.2020.111770

Pandi, S. R., Chari, N. V. H. K., Lotliker, A. A., Sarma, N. S., Tripathy, S. C., & Bajish, C. C. (2021a). Spatiotemporal variability in the optical characteristics of dissolved organic matter in the coastal Bay of Bengal. *International Journal of Environmental Science and Technology*. https://doi.org/10.1007/s13762-021-03605-7

Pandi, S. R., Chari, N. V. H. K., Sarma, N. S., Tripathy, S. C., Chiranjeevulu, G., & Das, S. (2021c). A review of estuarine CDOM dynamics of East Coast of India influenced by hydrographical forcing. In S. Das & T. Ghosh (Eds.), *Estuarine biogeochemical dynamics of the East Coast of India*. Springer. https://doi.org/10.1007/978-3-030-68980-3_14

Pandi, S. R., Chari, N. V. H. K., Sarma, N. S., Chiranjeevulu, G., Kiran, R., Murthy, K. N., Venkatesh, P., Lotliker, A. A., & Tripathy, S. C. (2021b). Characteristics of conservative and non-conservative CDOM of a tropical monsoonal estuary in relation to changing biogeochemistry. *Regional Studies in Marine Science, 44*, 101721. https://doi.org/10.1016/j.rsma.2021.101721

Pandi, S. R., Tripathy, S. C., Parida, C., Lotliker, A. A., Chandrasekhar Naik, R., Naik, R. K., Mishra, R. K., & Anilkumar, N. (2022). Spatiotemporal variability in bio-optical characteristics of the southwestern tropical Indian Ocean during boreal summer: Biophysical influences. *Progress in Oceanography, 208*, 102883. https://doi.org/10.1016/j.pocean.2022.102883

Pandi, S. R., Tripathy, S. C., Shaju, S. S., Minu, P., Kerkar, A. U., Bajish, C. C., & Anilkumar, N. (2023a). Interannual variability of surface bio-optical characteristics in the frontal zones of the Indian sector of the Southern Ocean during austral summer. *Polar Science*. https://doi.org/10.1016/j.polar.2023.100944

Pandi, S. R., Tripathy, S. C., Shaju, S. S., & Anilkumar, N. (2023b). Cross frontal variability in bio-optical characteristics in the Indian sector of the Southern Ocean during an austral summer. *Regional Studies in Marine Science, 61*, 102892. https://doi.org/10.1016/j.rsma.2023.102892

Pandi, S.R., Tripathy, S.C., Mendes, C.R.B., Naik, R.C., Padhi, S.K., Lotliker, A.A., & Mohan, R. (2025). Three-component models for the detection of phytoplankton size classes in the Indian sector of the Southern Ocean: Insights into climate-driven community shifts. *Journal of Marine Systems*, 251, 104115. https://doi.org/10.1016/j.jmarsys.2025.104115

Parida, C., Baliarsingh, S. K., Lotliker, A. A., Dash, M., Srichandan, S., & Sahu, K. C. (2019). Seasonal variation in optically active substances at a coastal site along western Bay of Bengal. *SN Applied Science,* 1, 1–8. https://doi.org/10.1007/s42452-019-1257-y

Patil, P. A., Arjun, A., & Menon, H. B. (2023). Bio-optical complexity and radiant heating rates in the coastal waters of eastern Arabian Sea. *Science of the Total Environment, 884*, 163838. https://doi.org/10.1016/j.scitotenv.2023.163838

Prakash, P., Prakash, S., Rahaman, H., Ravichandran, M., & Nayak, S. (2012). Is the trend in chlorophyll- a in the Arabian Sea decreasing ? *Geophysical Research Letters*, 39, 1–7. https://doi.org/10.1029/2012GL054187

Prasanna Kumar, S., Nuncio, M., Narvekar, J., Kumar, A., Sardesai, S., De Souza, S. N., Gauns, M., Ramaiah, N., & Madhupratap, M. (2004). Are eddies nature's trigger to enhance biological productivity in the Bay of Bengal? *Geophysical Research Letters*, 31, 2–6. https://doi.org/10.1029/2003GL019274

Preethi-Latha, T., Nagamani, P. V., Rao, B. S., Amarendra, P., Rao, K. H., Choudhury, S. B., et al. (2013). Particle backscattering variability in the coastal waters of bay of Bengal: A case study along off Kakinada and Yanam regions. *Geoscience and Remote Sensing Letters, 10*(6), 1517–1521.

Preethi-Latha, T., Nagamani, P. V., Rao, K. H., Dash, S. K., Choudhury, S. B., Rehman, A., et al. (2014). Retrieval and validation of chlorophyll-a concentrations in the coastal waters off Yanam and Kakinada (Godavari) basin along east coast of India. *Journal of the Indian Society of Remote Sensing, 42*, 129–138. https://doi.org/10.1007/s12524-013-0282-3

Qasim, S. Z. (1999). Productivity of the northern Indian Ocean. *Indian Journal of Marine Sciences, 28*, 85–91.

Rakesh, V., Singh, R., Yuliya, D., Pal, P.K. & Joshi, P.C. (2008). Impact of variational assimilation of MODIS thermodynamic profiles in the simulation of western disturbance. *International Journal of Remote Sensing, 30*(18), pp.4867–4887.

Rochelle-Newall, E. J., & Fisher, T. R. (2002). Chromophoric dissolved organic matter and dissolved organic carbon in Chesapeake Bay. *Marine Chemistry*, 77, 23–41. https://doi.org/10.1016/S0304-4203(01)00073-1

Rouf, M. A., Golder, M. R., & Sumana, Z. A. (2021). Satellite-based observation of particulate organic carbon in the northern bay of Bengal. *Environmental Advances, 6*, 100124. https://doi.org/10.1016/j.envadv.2021.100124

Roy, R., Chitari, R., Kulkarni, V., Krishna, M. S., Sarma, V. . V. S. S., & Anil, A. . C. (2015). CHEMTAX-derived phytoplankton community structure associated with temperature fronts in the northeastern Arabian Sea. *Journal of Marine Systems, 144*, 81–91. https://doi.org/10.1016/j.jmarsys.2014.11.009

Sadeghi, A., Dinter, T., Vountas, M., Taylor, B., Altenburg-Soppa, M., & Bracher, A. (2012). Remote sensing of coccolithophore blooms in selected oceanic regions using the PhytoDOAS method applied to hyperspectral satellite data. *Biogeosciences*, 9, 2127–2143. https://doi.org/10.5194/bg-9-2127-2012

Sahay, A., Mishra, M. K., Prakash, C., & Ajai. (2014). Absorption characteristics of ocean water in the Arabian Sea during winter bloom from in situ measurements and Oceansat 2 OCM and MODIS data. *International Journal of Remote Sensing, 35*(9), 2982–2995. https://doi.org/10.1080/01431161.2014.894659

Sahay, A., Ali, S. M., Gupta, A., & Goes, J. I. (2017). Ocean color satellite determinations of phytoplankton size class in the Arabian Sea during the winter monsoon. *Remote Sensing of Environment, 198*, 286–296. https://doi.org/10.1016/j.rse.2017.06.017

Sanyal, P., Ray, R., Paul, M., Gupta, V.K., Acharya, A., Bakshi, S., Jana, T.K. & Mukhopadhyay, S.K. (2020). Assessing the dynamics of dissolved organic matter (DOM) in the coastal environments dominated by mangroves, Indian Sundarbans. *Frontiers in Earth Science, 8*, p.218.

Sarangi, R. K., Chauhan, P., & Nayak, S. R. (2002). Vertical di_use attenuation coefficient (*Kd*) based optical classi_cation of IRS-P3 MOS-B satellite ocean colour data. *Proceedings of the Indian Academy of Sciences—Earth & Planetary Sciences, 111*(3), 237–245.

Sarma, N. S., Pandi, S. R., Chari, N. V. H. K., Chiranjeevulu, G., Rayaprolu, K., Shiva Krishna, K., Bhaskara Rao, D., Venkatesh, P., Charan Kumar, B., & Raman, A. V. (2018). Spectral modeling of estuarine colored dissolved organic matter. *Current Science*. https://doi.org/10.18520/cs/v114/i08/1762-1767

Sarma, V. V. S. S., Rao, D. N., Rajula, G. R., et al. (2019). Organic nutrients nupport nigh primary production in the bay of 662 Bengal. *Geophysical Research Letters, 46*, 6706–6715. https://doi.org/10.1029/2019GL082262

Sarma, N. S., Baliar, S. K., Pandi, S. R., Anandrao, A., & Samanta, A. (2022). Noctiluca blooms intensify when northwesterly winds complement northeasterlies in the northern Arabian Sea: Possible implications. *Oceanologia*. https://doi.org/10.1016/j.oceano.2022.06.004

Sarma, N. S., Bailar, S. K., Lotliker, A. A., Pandi, S. R., Samanta, A., & Suchismita, S. (2023a). Sea surface temperature and phytoplankton abundance as crucial proxies for green Noctiluca bloom monitoring in the northeastern Arabian Sea: A case study. *Ocean Science Journal, 58*. https://doi.org/10.1007/s12601-022-00096-6

Sarma, N. S., Chiranjeevulu, G., Pandi, S. R., Bhaskara, D., & Sarma, V. V. S. S. (2023b). Coupling between chromophoric dissolved organic matter and dissolved inorganic carbon in Indian estuaries. *Science of the Total Environment, 905*, 167120. https://doi.org/10.1016/j.scitotenv.2023.167120

Sasaki, H., Miyamura, T., Saitoh, S.I., Ishizaka, J. (2005). Seasonal variation of absorption by particles and colored dissolved organic matter (CDOM) in Funka Bay, southwestern Hokkaido, Japan. *Estuarine, Coastal and Shelf Sciences*, 64, 447–458. https://doi.org/10.1016/j.ecss.2005.03.008

Sathyendranath, S., Stuart, V., Irwin, B. D., Maass, H., Savidge, G., Gilpin, L., & Platt, T. (1999). Seasonal variations in bio-optical properties of phytoplankton in the Arabian Sea. *Deep Sea Research Part II: Topical Studies in Oceanography, 46*, 633–653.

Sathyendranath, S., Cota, G., Stuart, V., Maass, H., & Platt, T. (2001). Remote sensing of phytoplankton pigments: A comparison of empirical and theoretical approaches. *International Journal of Remote Sensing, 22*, 249–273.

Schmidtko S, Stramma L, & Visbeck M. (2017). Decline in global oceanic oxygen content during the past five decades. *Nature*, 542(7641):335–339. https://doi.org/10.1038/nature21399. PMID: 28202958.

Schwarz, J. N., Kowalczuk, P., Kaczmarek, S., Cota, G. F., Mitchell, B. G., Kahru, M., Chavez, F. P., Cunningham, A., McKee, D., Gege, P., Kishino, M., Phinney, D. A., & Raine, R. (2002). Two models for absorption by coloured dissolved organic matter (CDOM). *Oceanologia*, 44, 209–241.

Shaju, S.S., Prasad, S.C., P.S., V., Samal, A.K., Menon, N., M., N., Avarachen Mathew, K., Abdul Nazar, A.K., & George, G. (2019). Effect of Mariculture on bio-optical properties and water quality of Gulf of Mannar and Palk Bay. *Regional Studies in Marine Science*, 29, 100618. https://doi.org/10.1016/j.rsma.2019.100618

Shanmugam, P. (2011a). A new bio-optical algorithm for the remote sensing of algal blooms in complex ocean waters. *Journal of Geophysical Research: Oceans, 116*, 1–12.

Shanmugam, P. (2011b). New models for retrieving and partitioning the colored dissolved organic matter in the global ocean: Implications for remote sensing. *Remote Sensing of Environment, 115*, 1501–1521. https://doi.org/10.1016/j.rse.2011.02.009

Shanmugam, P., Suresh, M., & Sundarabalan, B. (2013). An innovative algorithm to detect and characterize ocean surface algal blooms. *IEEE Journal of Selected Topics in Applied Earth Observations and Remote Sensing, 6*, 1879–1892. https://doi.org/10.1109/JSTARS.2012.2227993

Shanmugam, P., Varunan, T., Jaiganesh, S. N. N., Sahay, A., Chauhan, P., et al. (2016). Optical assessment of colored dissolved organic matter and its related parameters in dynamic coastal water systems. *Estuarine, Coastal and Shelf Science, 175*, 126–136.

Shanmugam, P., He, X., Singh, R. K., & Varunan, T. (2018). A modern robust approach to remotely estimate chlorophyll in coastal and inland zones. *Advances in Space Research, 61*, 2491–2509. https://doi.org/10.1016/j.asr.2018.02.024

Shetye, S. R., Gouveia, A. D., Shenoi, S. S. C., et al. (1993). The western boundary current of the seasonal subtropical gyre in the Bay of Bengal. *Journal of Geophysical Research, 98*, 945–954. https://doi.org/10.1029/92JC02070

Shetye, S.R., Gouveia A.D., Shankar D., Shenoi S.S.C., Vinayachandran P.N., Sundar D., Michael G.S., & Nampoothiri G. (1996). Hydrography and circulation in the western Bay of Bengal during the northeast monsoon (1996). *Journal of Geophysical Research: Oceans*, 101 (C6), 95JC03307, 14011–14025. https://doi.org/10.1029/95JC03307

Shifrin, K. S. (1998). *Physical optics of ocean water*. Retrieved from https://books.google.co.in/books?id=fBBK2zAGPcYC

Sieburth, J. M. N., Smetacek, V., & Lenz, J. (1978). Pelagic ecosystem structure: Heterotrophic compartments of the plankton and their relationship to plankton size fractions. *Limnology and Oceanography, 23*. https://doi.org/10.4319/lo.1978.23.6.1256

Siegel, D. A., Buesseler, K. A., Doney, S. C., Sailley, S. F., Behrenfeld, M. J., & Boyd, P. W. (2014). Global assessment of ocean carbon export by combining satellite observations and food-web models. *Global Biogeochemical Cycles, 28*. https://doi.org/10.1002/2013GB004743

Singh, R. P., & Chaturvedi, P. (2010). *Comparison of chlorophyll concentration in the bay of Bengal and the Arabian Sea using IRS-P4 OCM and MODIS aqua* (pp. 334–340). NISCAIR-CSIR.

Slade, W. H., & Boss, E. (2015). Spectral attenuation and backscattering as indicators of average particle size. *Applied Optics, 54*, 7264. https://doi.org/10.1364/ao.54.007264

Sosik, H. M., & Mitchell, B. G. (1995). Light absorption by phytoplankton, photosynthetic pigments and detritus in the California current system. *Deep-Sea Research Part II, 42*, 1717–1748. https://doi.org/10.1016/0967-0637(95)00081-G

Stedmon, C. A., Markager, S., & Bro, R. (2003). Tracing dissolved organic matter in aquatic environments using a new approach to fluorescence spectroscopy. *Marine Chemistry, 82*, 239–254. https://doi.org/10.1016/S0304-6794203(03)00072-0

Stedmon, C. A., & Bro, R. (2008). Oceanography methods: Characterizing dissolved organic matter fluorescence with parallel factor analysis: a tutorial. *Limnology and Oceanography: Methods*, 6, 572–579. https://doi.org/10.1016/S0304-4203(03)00072-0

Stomp, M., Huisman, J., Stal, L. J., & Matthijs, H. C. P. (2007). Colorful niches of phototrophic microorganisms shaped by vibrations of the water molecule. *The ISME Journal, 1*, 271–282. https://doi.org/10.1038/ismej.2007.59

Stramski, D., Bricaud, A., & Morel, A. (2001). Modeling the inherent optical properties of the ocean based on the detailed composition of the planktonic community. *Applied Optics*, 40, 2929–2945.

Stramski, D., Reynolds, R. A., Babin, M., Kaczmarek, S., Lewis, M. R., Röttgers, R., & Claustre, H. (2008). Relationships between the surface concentration of particulate organic carbon and optical properties in the eastern south Pacifc and eastern Atlantic oceans. *Biogeosciences, 5*, 171–201. https://doi.org/10.5194/bg-5-171-2008

Stramski, D., Joshi, I., & Reynolds, R. A. (2022). Ocean color algorithms to estimate the concentration of particulate organic carbon in surface waters of the global ocean in support of a long-term data record from multiple satellite missions. *Remote Sensing of Environment, 269*, 112776.

Sun, X., Brewin, R. J. W., Sathyendranath, S., et al. (2023). Coupling ecological concepts with an ocean-colour model: Phytoplankton size structure. *Remote Sensing of Environment,* 285, 113415. https://doi.org/10.1016/j.rse.2022.113415

Sunanda, N., Kuttippurath, J., Chakraborty, A., & Peter, R. (2023). Stressors of primary productivity in the north Indian Ocean revealed by satellite, reanalysis and CMIP6 data. *Progress in Oceanography,* 219, 103164. https://doi.org/10.1016/j.pocean.2023.103164

Suzuki, R., & Ishimaru, T. (1990). An improved method for the determination of phytoplankton chlorophyll using N,N-Dimethylformamide. *Journal of the Oceanographical Society of Japan, 46*, 190–194.

Tassan, S., & Ferrari, G. M. (1995a). An alternative approach to absorption measurements of aquatic particles retained on filters. *Limnology and Oceanography, 40*(8), 1358–1368. https://doi.org/10.4319/lo.1995.40.8.1358

Tassan, S., & Ferrari, G. M. (1995b). Proposal for the measurement of backward and total scattering by mineral particles suspended in water. *Applied Optics, 34*, 8345–8353. https://doi.org/10.1364/AO.34.008345

Tilstone, G. H., Angel-Benavides, I. M., Pradhan, Y., Shutler, J. D., Groom, S., & Sathyendranath, S. (2011). An assessment of chlorophyll-a algorithms available for SeaWiFS in coastal and open areas of the bay of Bengal and Arabian Sea. *Remote Sensing of Environment, 115*, 2277–2291. https://doi.org/10.1016/j.rse.2011.04.028

Tilstone, G.H., Lotliker, A.A., Miller, P.I., Ashraf, P.M., Kumar, T.S., Suresh, T., Ragavan, B.R., & Menon, H.B. (2013). Assessment of MODIS-Aqua chlorophyll-a algorithms in coastal and shelf waters of the eastern Arabian Sea. *Continental Shelf Research*, 65, 14–26. https://doi.org/10.1016/j.csr.2013.06.003

Tilstone, G. H., Xie, Y., Robinson, C., Serret, P., Raitsos, D. E., Powell, T., Aranguren-gassis, M., Garcia-martin, E. E., & Kitidis, V. (2015). Satellite estimates of net community production indicate predominance of net autotrophy in the Atlantic Ocean. *Remote Sensing of Environment*, 164, 254–269. https://doi.org/10.1016/j.rse.2015.03.017

Tiwari, S.P. & Kolluru, S., (2023) Understanding Primary Productivity in the Indian Ocean Using Bio-Optics and Remote Sensing. In *Dynamics of Planktonic Primary Productivity in the Indian Ocean* (pp. 207–244). Cham: Springer International Publishing.

Tiwari, S. P., & Shanmugam, P. (2012). An optical model for the remote sensing of absorption coefficients of phytoplankton in oceanic/coastal waters. *Advanced Remote Sensing, 1*(2), 19–34.

Tiwari, S. P., & Shanmugam, P. (2013). An evaluation of models for the satellite-estimation of phytoplankton absorption coefficients in coastal/oceanic waters. *IEEE Journal of Selected Topics in Applied Earth Observations and Remote Sensing, 7*(1), 364–371.

Tomkins, M., Martin, A. P., Nurser, A. J. G., & Anderson, T. R. (2020). Phytoplankton acclimation to changing light intensity in a turbulent mixed layer: A Lagrangian modelling study. *Ecological Modelling, 417*, 108917.

Tripathy, S. C., Sabu, P., Patra, S., Naik, R. K., Sarkar, A., Venkataramana, V., Kerkar, A. U., & Pandi, S. R. (2020). Biophysical control on variability in phytoplankton production and composition in the South-Western tropical Indian Ocean during monsoon 2014. *Frontiers in Marine Science, 7*, 515. https://doi.org/10.3389/fmars.2020.00515

Uitz, J., Claustre, H., Morel, A., & Hooker, S. B. (2006). Vertical distribution of phytoplankton communities in open ocean: An assessment based on surface chlorophyll. *Journal of Geophysical Research, Oceans, 111*. https://doi.org/10.1029/2005JC003207

Vantrepotte, V., Loisel, H., Mélin, F., Desailly, D., & Gaurier, L. D. (2011). Global particulate matter pool temporal variability over the Seawifs period (1997–2007). *Geophysical Research Letters, 38*(2), 1–5. https://doi.org/10.1029/2010GL046167

Varunan, T., & Shanmugam, P. (2015). A model for estimating size-fractioned phytoplankton absorption coefficients in coastal and oceanic waters from satellite data. *Remote Sensing of Environment, 158*, 235–254.

Vidussi, F., Claustre, H., Manca, B. B., Luchetta, A., & Jean-Claude, M. (2001). Phytoplankton pigment distribution in relation to upperthermocline circulation in the eastern Mediterranean Sea during winter. *Journal of Geophysical Research*, 106, 939–956.

Vinayachandran, P. N. M., Masumoto, Y., Roberts, M. J., Huggett, J. A., Halo, I., Chatterjee, A., Amol, P., Gupta, G. V. M., Singh, A., Mukherjee, A., Prakash, S., Beckley, L. E., Raes, E. J., & Hood, R. (2021). Reviews and syntheses: physical and biogeochemical processes associated with upwelling in the indian ocean. *Biogeosciences*, 18, 5967–6029. https://doi.org/10.5194/bg-18-5967-2021

Vishnu, P. S., Shaju, S. S., Tiwari, S. P., Menon, N., Nashad, M., Joseph, C. A., & Mohandas, A. (2018). Seasonal variability in bio-optical properties along the coastal waters of Cochin. *International Journal of Applied Earth Observation and Geoinformation, 66*, 184–195.

Wright, S. W., & Jeffrey, S. W. (2006). Pigment markers for phytoplankton production. In: Volkman, J.K. (eds).

Xi, H., Losa, S. N., Mangin, A., Soppa, M. A., Garnesson, P., Demaria, J., Liu, Y., Hembise, O., & Bracher, A. (2020). Remote Sensing of Environment Global retrieval of phytoplankton functional types based on empirical orthogonal functions using CMEMS GlobColour merged products and further extension to OLCI data. *Remote Sensing of Environment,* 240, 111704. https://doi.org/10.1016/j.rse.2020.111704

Xing, X., Claustre, H., Uitz, J., Mignot, A., Poteau, A., & Wang, H. (2014). Seasonal variations of bio-optical properties and their interrelationships observed by bio-Argo floats in the subpolar North Atlantic. *Journal of Geophysical Research, Oceans, 119*, 7372–7388. https://doi.org/10.1002/2014JC010189

Yang, L., Hong, H., Chen, C.A., Guo, W., & Huang, T. (2013). Chromophoric dissolved organic matter in the estuaries of populated and mountainous Taiwan. *Marine Chemistry*, 157, 12–23. https://doi.org/10.1016/j.marchem.2013.07.002

Yu, X., Salama, M. S., Shen, F., & Verhoef, W. (2016). Retrieval of the diffuse attenuation coefficient from GOCI images using the 2 SeaColor model: A case study in the Yangtze estuary. *Remote Sensing of Environment, 175*, 109.

Zhang, T., & Fell, F. (2007). An empirical algorithm for determining the diffuse attenuation coefficient Kd in clear and turbid waters from spectral remote sensing reflectance. *Limnology and Oceanography: Methods*, 5, 457–462. https://doi.org/10.4319/lom.2007.5.457

Zhao, W., Wang, J., & Chen, M. (2009). Three-dimensional fluorescence characteristics of dissolved organic matter produced by Prorocentrum donghaiense Lu. *Chinese Journal of Oceanology and Limnology*, 27, 564–569. https://doi.org/10.1007/s00343-009-9141-z

Chapter 4
Optical Characterization and Remote Sensing of Phytoplankton Dynamics in the South Eastern Arabian Sea: Insights and Applications

P. Minu, V. P. Souda, S. S. Shaju, Aneesh A. Lotliker, and P. Muhamed Ashraf

Abstract This book chapter provides a comprehensive overview of optical characterization and remote sensing applications in understanding phytoplankton dynamics and associated environmental factors in the coastal waters of the South Eastern Arabian Sea (SEAS). Through multiple studies conducted over various seasons from 2010 to 2017, the research investigates the abundance, composition, and optical properties of phytoplankton, as well as the influence of environmental parameters such as nutrient concentrations and seasonal variations on their distribution. Key findings include the identification of diverse phytoplankton genera, predominantly centric diatoms, with notable seasonal variations in abundance, diversity, and pigment composition. The research highlights the influence of environmental factors such as nitrate, phosphate, and silicate concentrations on phytoplankton size distribution, emphasizing the importance of understanding these dynamics for accurate remote sensing predictions. Furthermore, the chapter delves into the role of colored dissolved organic matter (CDOM) in modulating underwater light penetration and its temporal variations in coastal waters. The study reveals CDOM sources and their implications for optical complexity, emphasizing the need to incorporate seasonal components and phytoplankton abundance in CDOM algorithms. Additionally, the chapter explores the optical classification of coastal waters based on spectral reflectance and optically active substances (OAS), including chlorophyll-a (Chl-a), CDOM, and total suspended matter (TSM). Evaluation of bio-

P. Minu (✉)
Department of Aquatic Environment Management, Kerala University of Fisheries and Ocean Studies, Ernakulam, Kerala, India

V. P. Souda · P. M. Ashraf
Fishing Technology Division, CIFT (ICAR Institute), Kochi, Kerala, India

S. S. Shaju
Department of Chemical Oceanography, CUSAT, Kochi, Kerala, India

A. A. Lotliker
Indian National Centre for Ocean Information Services, Ministry of Earth Sciences, Hyderabad, Telangana, India

© The Author(s), under exclusive license to Springer Nature Switzerland AG 2026

S. Das (ed.), *Bio-optical Properties of Indian Coastal Waters*,
https://doi.org/10.1007/978-3-032-08714-0_4

optical algorithms for Chl-a retrieval from satellite data underscores the significance of atmospheric correction schemes in enhancing accuracy, with insights into the impact of environmental variables on remote sensing performance. Overall, this chapter contributes to a deeper understanding of phytoplankton dynamics, optical properties, and their implications for remote sensing applications in coastal ecosystems, providing valuable insights for both scientific research and practical management of marine resources.

Keywords Phytoplankton dynamics · Remote sensing · Optical properties · Colored dissolved organic matter (CDOM) · South Eastern Arabian Sea (SEAS)

1 Introduction

Observations of ocean color obtained through remote sensing can be effectively utilized for the prediction of global dynamics of marine ecosystems. Currently, Chlorophyll-a (Chl-a) is regarded as the primary biological factor influencing ocean color. Additional details on phytoplankton taxonomy and pigment composition can enhance such predictions. Links between optical properties, pigment composition of phytoplankton, and other photosynthetic parameters have been reported recently (Morel, 1997; Kahru & Mitchell, 1998; Stuart et al., 1998; Stramski et al., 2001; Aiken et al., 2004; Barlow et al., 2002, 2004; Moore et al., 2005; Fishwick et al., 2006). But these links are seriously affected by phytoplankton populations encompassing species that vary in pigment composition and size structure. Optical properties of surface waters, affected by variability in phytoplankton communities, can be described by identifying factors that are co-varying with Chl-a (Ciotti et al., 2002). Absorption spectra for phytoplankton communities can be distinguished based on taxonomic composition and cell size, and the implications of these various communities on the relationship between optical properties can be investigated.

2 Methodology

Several locations in the coastal waters near Kochi were used to analyze the monthly bio-optical features of water samples. Figure 4.1 shows the locations of the sampling stations. The study region is a portion of the South Eastern Arabian Sea (SEAS) and undergoes considerable seasonal fluctuations in hydrographic conditions due to the reversal of monsoons. This influences surface circulation and river water discharge in the area. The upwelling effect promotes the southerly current in the coastal waters during the monsoon season (Joshi & Rao, 2012). During the monsoon season, nutrient concentrations in coastal waters increase, while both surface salinity and water temperatures decrease. The amount of detritus increases, resulting in diminished light penetration. These changes help in higher production

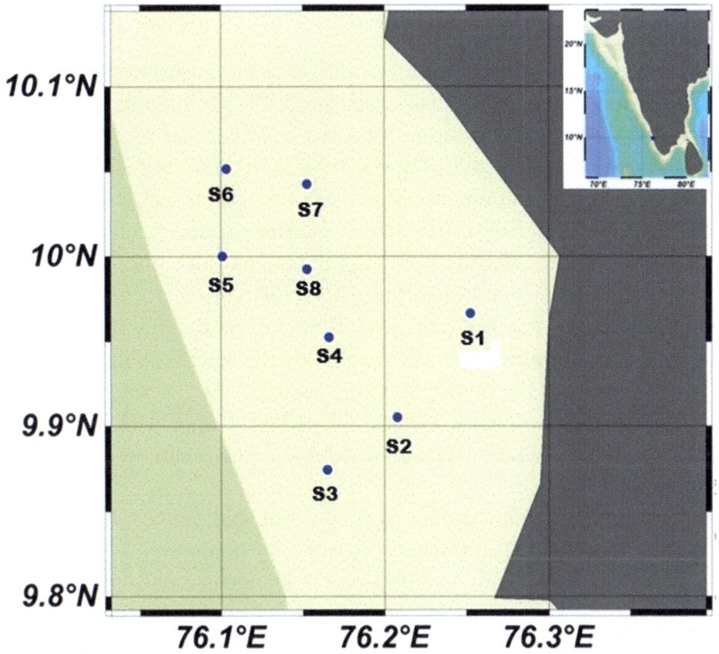

Fig. 4.1 Study area map showing the eight coastal stations in the SEAS

at both primary and secondary levels (Devassy & Goes, 1989; Madhupratap et al., 1990).

At certain locations during the southwest monsoon period, seasonal phytoplankton blooms (Srinivas & Kumar, 2006) were also observed. Bhagirathan et al. (2014) pointed out that an increase in the influx of nutrients contributes to the domination of a single species of phytoplankton and results in a high pelagic fishery. The southwest monsoon started on June first and continued until October 21st, 2013. Over the region enclosed by latitudes 5–10° N and longitudes 70–80° E, the average wind speed at 925 hPa was 19 knots. The year 2013 was a non-El Niño year. Data indicated that continuous and intense rainfall throughout the study region (considering Kerala in its entirety) led to an unusual excess of rainfall, measuring +61% more than the normal levels in June.

Station 4 represents the area where mixing of marine and freshwater occurs, and Station 1 is in the nearby estuarine area, from which increased nutrient-rich freshwater influx occurs to the adjacent coastal waters during the monsoon season. Other stations were located on the northern and southern portions of an inlet to the Kochi estuary.

Using a commercial purse seiner, surface water samples were collected monthly from eight sampling sites between May 2013 and January 2014. The samples were collected using a Niskin water sampler and a clean bucket (for surface samples), stored in clean, dark polyethylene bottles, kept cold, and transported to the laboratory for further examination.

The absorption by particulate matter (aph) and the absorption by colored dissolved organic matter (aCDOM) were measured 48 h following the collection of samples and 8 h of sample collection, respectively. Water was passed through nitrocellulose membrane filters (with a pore size of 0.2 μm), and the resulting filtrate was utilized for absorption measurements (Kowalczuk & Kaczmarek, 1996; Twardowski et al., 2004; Minu et al., 2020). For aph(λ) measurements, the filters were rinsed with hot methanol before analysis following the quantitative filter pad method.

The difference between the absorption of total particulate matter (ap(λ)) and non-pigmented detritus material (ad(λ)) is calculated as the phytoplankton absorption (aph(λ). Water samples were taken individually and kept in a formaldehyde (4%) solution to identify the phytoplankton species. Cell L-1 was used to measure the total phytoplankton density (Utermohl, 1958; Cupp, 1943; Tomas, 1997; Subrahmanyan, 1959). Small-sized phytoplankton fractionation was not a part of the procedure.

At every station, a stand-alone hyperspectral radiometer from Satlantic™ Instruments was deployed. The HyperOCR is a fully functional, stand-alone device that can produce light spectrum data obtained in an ocean environment. The instrument contains various sensors for determining ambient above-water irradiance and radiance. In addition to the primary sensors, it has supplementary sensors to measure pressure, temperature, water column conductivity, and the instrument's tilt. It also has an ECOtriplet sensor (Wetlabs ECO series), which detects chlorophyll fluorescence and CDOM fluorescence. The radiometer operated three times, away from the boat's shadow. The data from the hyperspectral radiometer was recorded using SatView™ software, and multi-level processing was carried out using Prosoft™ software (Fargion & Mueller, 2000).

3 Phytoplankton Diversity, Numerical Abundance, and Seasonal Variation

With 73 genera spread across 19 orders and 41 families, the phytoplankton in Kochi's coastal waters has a rich taxonomic composition. The region was dominated by the order Biddulphiales, family Thalassiosiraceae and *Skeletonema* spp. dominated the area. The spatial and temporal patterns exhibited a shifting community composition, where diatoms predominantly occupied the ecosystem, changing to dinoflagellates and cyanobacteria. The highest diversity occurred along the outer Barmouth region. The diversity was lower in the estuarine areas. Diversity was low during the southwest monsoon and higher during the pre-monsoon season. The Shannon-Weiner diversity index indicated that the highest diversity levels were observed after the monsoon, while diversity was low during the monsoon season. These results not only shed light on the complex composition and dynamics of phytoplankton communities in coastal ecosystems but also highlight how environmental conditions affect their spatial and temporal variations (Minu et al., 2014) (Table 4.1) (Fig. 4.2).

In coastal ecosystems, phytoplankton abundance exhibits complex responses to seasonal and environmental changes. During the monsoon season, there is a notable decrease in the numbers of different phytoplankton groups, which is linked to several contributing factors. These factors include reduced pH and temperature, along with increased nitrite and silicate levels, combined with lower phosphate concentrations (Shaju et al., 2015). Enhanced rainfall during the monsoon causes an increase in freshwater flow, which alters the physicochemical characteristics of coastal waters and creates adverse conditions for the growth of phytoplankton. Additionally, the influx of land-derived nutrients due to runoff further alters the nutrient imbalance, hindering phytoplankton development. However, as the ecosystem shifts into the pre-monsoon period, there is a decrease in river runoff, leading to better water clarity and significant reductions in nitrate and silicate concentrations. The reduction in nutrient availability creates a more favorable environment for the growth of phytoplankton, leading to greater species diversity in coastal waters. Research has consistently observed this occurrence, underscoring the role of nutrient dynamics in influencing phytoplankton community structure and seasonal fluctuations in coastal regions. Understanding these complex responses is crucial for the effective

Table 4.1 List of phytoplankton species identified during the study

1. *Diatoms*
Centric diatoms: *Skeletonema costatum, Cyclotella, Lauderia annulata, Planktoniella* spp., *Thalassiosira subtilis, Melosira, Leptocylindrus danicus, Corethron* spp., *Coscinodiscus* spp., *Hemidiscus, Azpeitia, Asteromphalus, Guinardia, Eucampia* spp., *Hemiaulus, Triceratium, Ditylum brightwellii, Streptotheca, Odontella, Climacosphenia, Licmophora, Cyclophora, Rhizosolenia* spp.: *R. setigera, R. delicatula, R. imbricate, R. styliformis,*
R. stolterfothii, R. alata, Bacteriastrum spp.: *B. hyalinum, B. delicatulam, B. furcatum,*
Chaetoceros spp.: *C. decipiens, C. curvisetus, Biddulphia* spp.: *Biddulphia sinensis, B. mobiliensis, B. aurita*
Pennate diatoms: *Phaeodactylum, Striatella, Thalassionema* spp., *Amphiprora, Asterionella japonica, Cylindrotheca closterium, Synedra, Nanoneis, Pseudoeunotia, Fragilaria, Thalassiothrix, Navicula* spp., *Pseudonitzschia* spp., *Pleurosigma* spp.: *Pleurosigma elongatum, P. directum, Nitzschia* spp.: *N. seriata, N. sigma, N. longissima*
2. *Dinoflagellates*
Mesoporos, Gymnodinium, Gyrodinium, Cochlodinium, Katodinium, Pyrocystis, Pyrophacus, Alexandrium, Gonyaulax, Lingulodinium, Peridinium, Preperidinium, Scrippsiella, Noctiluca, Heterocapsa, Prorocentrum **spp.:** *P. micans, P. lima*
Dinophysis **spp.:** *D. caudata, D. acuminate, D. fortii, D. tripos, Ceratium* **spp.:** *Ceratiumfusus, C. tripos, C. furca, C. inflatum, C. horridum, Protoperidinium* **sp.:** *P. ovatum, P. leonis*
3. *Green algae*
Agmenellum, Oscillatoria, Lyngbya, Anabaena, Nostoc, Staurastrum
4. *Blue-green algae*: *Trichodesmium erythraeum, synechococcus*
5. *Phyto flagellates: Euglena, Pediastrum* spp., *Scenedesmus, Actinastrum, Dictyocha*

Source: Minu, 2017

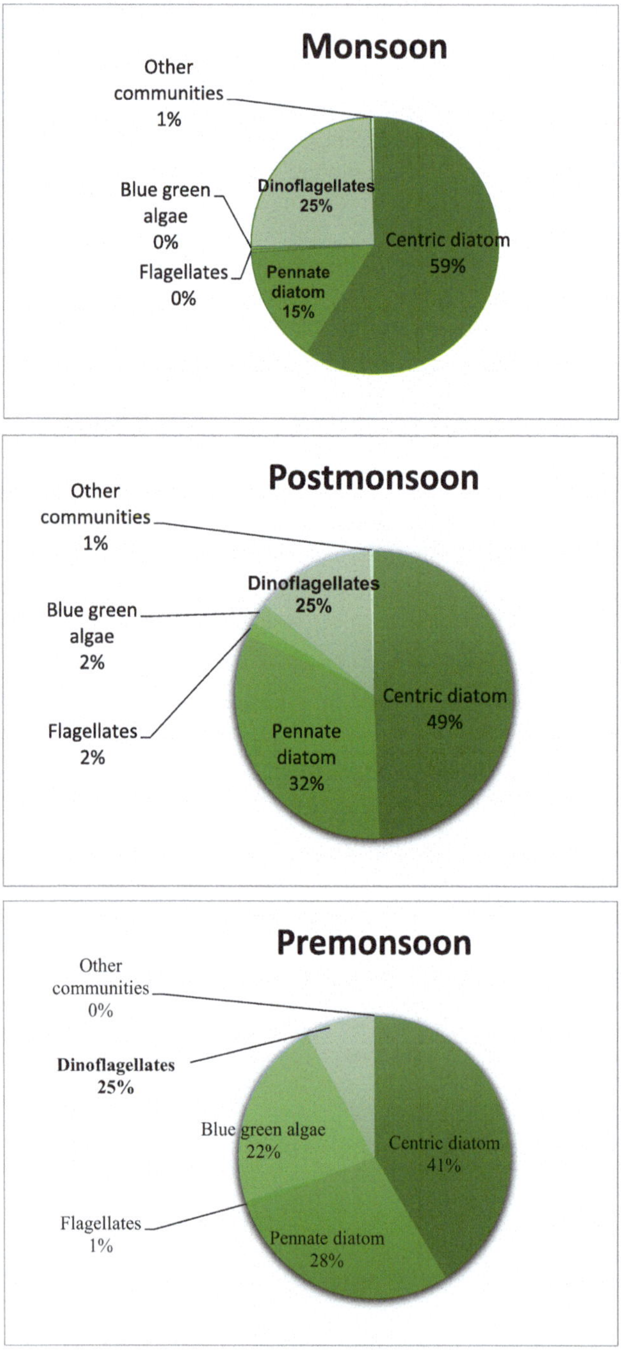

Fig. 4.2 Phytoplankton community percentage compositions in the study area during the course of three seasons

management and conservation of ecosystems in coastal areas, underscoring the need for integrated strategies that include both natural and human-induced factors affecting ecosystem dynamics. Solar radiation, salinity, and temperature changes during this phase also promote the proliferation of blue-green algae, especially Trichodesmium spp., which aligns with the identified relationships between phytoplankton abundance and environmental conditions.

4 Bio-Optical Characteristics Along the Kochi Coast

4.1 Chlorophyll Concentration

Chlorophyll-a (Chl-a) levels in coastal waters act as an essential measure of primary productivity and the overall health of ecosystems. As the main pigment found in photosynthetic organisms such as phytoplankton, Chl-a is crucial in the marine food web, serving as the foundation of the trophic pyramid. The concentration of Chl-a reflects both the abundance and activity of phytoplankton, which are fundamental to marine food chains and support a wide variety of aquatic life, including zooplankton and fish. Additionally, factors such as nutrient availability, light conditions, water temperature, and hydrodynamic forces impact Chl-a levels, making it a sensitive marker of environmental fluctuations and ecological dynamics in coastal habitats. Understanding the dynamics of Chl-a concentrations aids in monitoring, management, and conservation initiatives by offering crucial insights into the productivity and health of coastal marine ecosystems.

Throughout the duration of the study, the average chlorophyll concentration fluctuated from 1.375 ± 1.661 to 10.139 ± 9.486 mg m^{-3}. The estuarine region experienced high chlorophyll concentration (Min: 1.103; Max: 39 mg m^{-3}), followed by the Barmouth region (Min: 0.525; Max: 38.4 mg m^{-3}). Annual variability analysis revealed a wave periodicity of 5 years for Chl-a from 2008 to 2017 (Souda et al., 2020). The coastal waters experienced a minimum chlorophyll of 0.0277 mg m^{-3} and a maximum of 32.566 mg m^{-3}. There was not much difference in the chlorophyll concentration of surface waters and the 10 m depth of coastal stations. At the 20 m depth of coastal stations, the higher frequency of chlorophyll concentration was between 0.5 and 5 mg m^{-3}.

The intra-annual variability of chlorophyll concentration in the study area is shown in Fig. 4.3. In January, the chlorophyll concentration in the area was low, with a slightly elevated concentration observed in the upper part of the estuarine region. This pattern occurred both in 2015 and 2016. During February 2014, the chlorophyll concentration decreased as one proceeded away from the estuary to offshore waters. February in both 2015 and 2016 followed the same pattern as January. In February and April of 2015, the concentration of Chl-a was notably higher in the 10 m depth water column within the longitudes of 76.18° to 76.16° E, whereas the remainder of the region showed low Chl-a levels (2 mg m^{-3}). A distinct

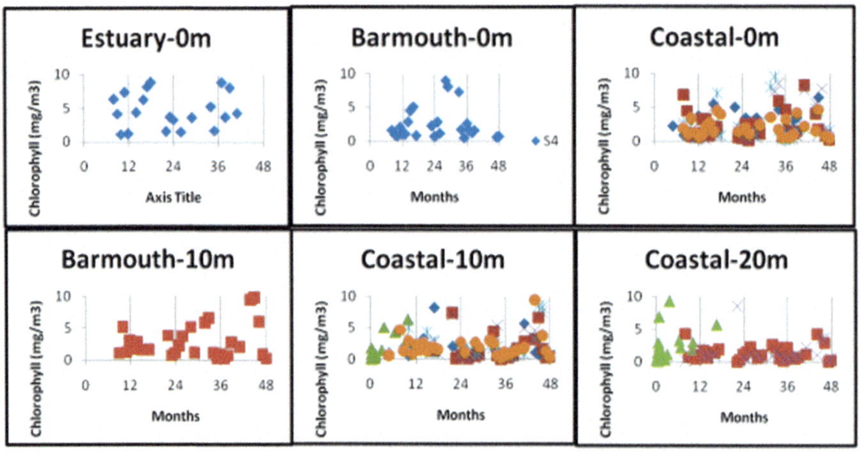

Fig. 4.3 Chlorophyll concentration for surface waters at 10 m depth and 20 m depth

stratification of Chl-a was noted in August and September across the entire study period. Surface waters were with very high Chl-a (15–20 mg m^{-3}), 5–10 m water column with 6–10 mg m-3 of Chl-a concentration, and bottom waters with low Chl-a (< 5 mg m^{-3}). During this period, the 10 m water column exhibited elevated Chl-a levels between longitudes 76.18° and 76.16° E. The stratification also pertained during October also. November and December exhibited low Chl-a in the area.

The chlorophyll concentration for different size fractions was also studied. The lowest concentration of Chl_{micro} recorded throughout the study was 0.00173 mg m^{-3}, while the highest reached 20.66 mg m^{-3}. The average concentration of Chl_{micro} in the estuarine area was 4.314 ± 5.206 mg m^{-3}, whereas in Barmouth, it measured 2.555 ± 4.821 mg m^{-3}. The concentrations of Chl_{micro} at coastal stations ranged from 0.0226 ± 0.048 mg m^{-3} at a depth of 10 m to 0.014 ± 0.041 mg m^{-3} at a depth of 20 m. The average Chl_{nano} measured in the estuarine station was 4.672 ± 3.93 mg m^{-3} with a minimum of 0.415 mg m^{-3} and a maximum of 15.36 mg m^{-3}. The Barmouth station had an average Chl_{nano} 2.187 ± 3.513 mg m^{-3} and experienced a minimum of 0.010 mg m^{-3} and a maximum of 18.88 mg m^{-3}. The minimum and maximum concentrations of Chl_{nano} observed at the 10 m bathymetric stations were 0.0082 mg m^{-3} and 0.11 mg m^{-3}, respectively. The 20 m bathymetric stations had Chl_{nano} ranging from 0.003 mg m^{-3} to 0.043 mg m^{-3}. The average chlorophyll concentration of picoplankton (Chl_{pico}) in the estuarine, 10 m bathymetric stations and 20 m bathymetric stations was 0.508 ± 0.663 mg m^{-3}, 0.384 ± 0.5607 mg m^{-3} and 0.305 ± 0.476 mg m^{-3}, respectively, in the surface waters.

Seasonal variations further influenced the Chl-a dynamics, with higher concentrations during specific months at different depths, implicating the interplay of environmental factors such as light availability and nutrient dynamics. The study elucidated the role of phytoplankton in driving these variations, underscoring the packaging effect of absorption, particularly pronounced in shallower regions. Furthermore, the influence of larger climate cycles, such as La Nina/El Nino events occurring every 5–6 years, was evident, shaping Chl-a concentration dynamics in a sinusoidal pattern. These underscore the intricate interplay between local and large-scale environmental factors in governing coastal Chl-a levels, highlighting its significance as a key indicator of ecosystem health and productivity in coastal waters.

4.2 Spectral Variation in Remote Sensing Reflectance

Remote sensing reflectance is calculated by dividing the upwelling radiance that emerges from the ocean by the downwelling irradiance that reaches the water's surface. It is proportional to the ratio of backscattering (bb) and absorption (a), coefficients, [bb/(a + bb)] (Morel & Prieur, 1977). A hyperspectral radiometer was used to measure the remote sensing reflectance.

To calculate the spectral Rrs,

$$\mathrm{Rrs}\left(0^{+},\lambda\right)=\frac{\mathrm{Lw}\left(0^{+},\lambda\right)}{\mathrm{Ed}\left(0^{+},\lambda\right)} \tag{4.1}$$

where Ed (0+, λ) is downwelling irradiance above the sea surface and Lw (0+, λ) is water-leaving radiance. Additionally, the following formula was used to calculate the upwelling and downwelling radiances.

$$\mathrm{Ed}\left(0^{+},\lambda\right)=\frac{\mathrm{Ed}\left(0^{-}\lambda\right)}{1-\alpha} \tag{4.2}$$

where

α—Fresnel reflection albedo for irradiance from sun and sky

and

$$\mathrm{Lw}\left(0^{+},\lambda\right)=\mathrm{Lu}\left(0^{-},\lambda\right)\frac{\left[1-\rho\left(\lambda,\theta\right)\right]}{\eta_{w}^{2}\left(\lambda\right)} \tag{4.3}$$

where

$\rho(\lambda,\theta)$—Fresnel reflectance index of seawater;
$\eta w(\lambda)$—Fresnel refractive index of seawater.

Three different types of spectra were identified from the Rrs data obtained in the study area using a hyperspectral radiometer; Fig. 4.4 shows the spectral diversity of these spectra. The coastal waters close to Kochi can be optically divided into three classes based on variations in Rrs spectra. The spectra of Type-I (T1) waters showed a smoother curve from 400 to 450 nm, followed by a rise to a peak at 482 nm (Fig. 4.4a). Rrs gradually decreased after 482 nm until it reached 608 nm, at which

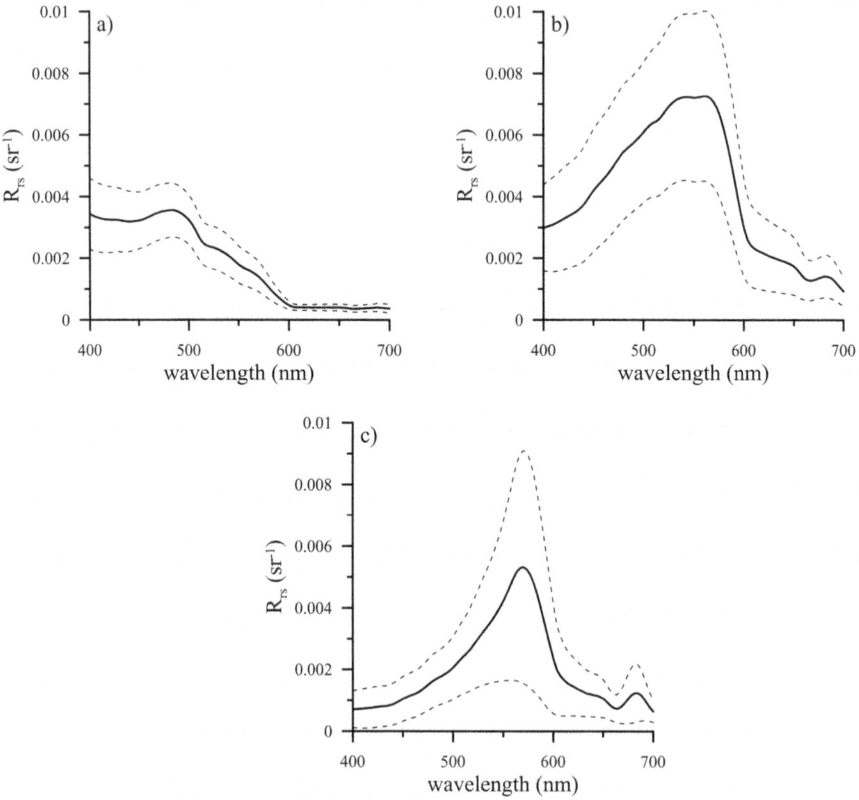

Fig. 4.4 Mean and standard deviation of three classes of reflectance spectra acquired during the study. The *X* axis represents wavelength (nm) and the *Y* axis represents the coefficient of reflectance (Sr^{-1}). (Source: Minu et al., 2016)

point the curve leveled off once more until it reached 700 nm, at which point there was no reflectance. It was found that the spectral Rrs in Type-II (TII) waters differed from those in T1. From a shorter wavelength (400 nm), the average Rrs increased gradually. It had a nearly flat peak between 532 and 566 nm, with a slightly larger value at 560 nm. A sharp decline was seen beyond 560 nm, reaching 610 nm. The reduction then proceeded gradually until it reached 670 nm. At 681 nm, there was also a secondary maximum. At longer wavelengths, the Rrs spectra of Type-III (TIII) waters resembled those of Type-II. Rrs increased sharply in this range between 400 and 570 nm. The Rrs peak was more noticeable. In Type-III, the spectral behavior of Rrs from 570 nm to 700 nm was comparable to Type-II. At 684 nm, however, the secondary peak was more noticeable and had a significant magnitude (Table 4.2).

Chlorophyll, total suspended matter, and CDOM are three optically active substances whose contributions were used to assess the cause of these three forms of Rrs. There are significant differences in the distribution of optically active substances (OAS) in different types of water, each manifesting distinct spectral

Table 4.2 Mean and standard deviation of three classes of reflectance spectra obtained during the study

	T1	T2	T3
Chl-a (mg m^{-3})	0.25(±0.17)	7.4 (± 4.8)	11.5 (± 6.8)
β_{650} (m^{-1} sr^{-1})	0.3 (±0.1) × 10^{-3}	2.8 (± 1.7) × 10^{-3}	3.2 (± 1.7) × 10^{-3}
CDOM (ppb QSDE)	0.94 (± 0.5)	1.69 (± 0.5)	2.23 (±0.37)

Source: Minu et al., 2016

signatures in remote sensing reflectance (Rrs) spectra. The Type-I waters, which are distinguished by little variation in the concentrations of colored dissolved organic matter (CDOM) and chlorophyll-a (Chl-*a*), Rrs spectra show a slight presence in the red area (beyond 600 nm) and a peak in the blue band (400 to 480 nm). The combined effect of Chl-*a* and CDOM absorption dominates the spectral signature in Type-II waters, resulting in different Rrs features. This is why the Rrs spectra in these waters exhibit a noticeable difference in the shorter wavelengths. Compared to Type-II waters, Type-III waters show a more noticeable peak in their Rrs spectra and a further shift towards longer wavelengths. Chl-*a* and CDOM concentrations increase by 64% and 76%, respectively, whereas β650 stays similar to Type-II waters.

4.3 Spectral Variation in Absorption by Colored Dissolved Organic Matter

The presence of CDOM in marine environments primarily arises from the breakdown of biological processes and inputs originating from land, like river runoff. The research highlighted that CDOM derives from terrigenous sources during the monsoon season, while in situ production contributes during the pre-monsoon and post-monsoon periods from both autochthonous and allochthonous sources (Minu et al., 2020). The range of the CDOM (QSDE) frequency distribution was 0.39 to 2.75 ppb L^{-1}. April 2009 had the highest recorded value of 2.75 ppb L^{-1}, while January 2011 had the lowest value of 0.39 ppb L^{-1}. Concentrations ranging from 0.8 to 2.4 ppb L^{-1} showed high frequency. The study also found that the reason for the unusually high CDOM in the region in January and February was the deteriorating bloom of Noctiluca scintillans.

4.4 Spectral Variation in Absorption by Detrital Matter

Absorption by the detritus fraction showed that it is related to both bathymetry and horizontal distance from the shore. The absorption by detritus decreased with a greater distance from the coast, while absorption rose with increasing depth. In the coastal waters off Kochi, detritus is recognized as the primary light absorber, followed by phytoplankton and CDOM, as demonstrated by both in situ data and OC-CCI-derived information.

4.5 Spectral Variation in Absorption by Phytoplankton Absorption

Absorption by total phytoplankton and different size classes, viz. micro, nano, and pico phytoplankton. The findings indicated that microphytoplankton had the highest absorption during the monsoon season. Absorption by nanophytoplankton at 443 nm varied greatly in the surface waters. The absorption by total phytoplankton is influenced by the bathymetry of the region, i.e., as depth increases, absorption decreases. Samples from the north of the Barmouth region were found to have higher absorption compared to the southern part. Analysis of annual variability indicated a 5-year periodic wave for Chl-a between 2008 and 2017. The phytoplankton absorption declined from 2013 to 2017. Spatial variations due to absorption by total phytoplankton and nanophytoplankton decreased with increasing distance offshore. Further, vertical differences, driven by the trade winds from the southwest and northeast, were significant for the absorption by nanophytoplankton during the monsoon and post-monsoon seasons. Stations along 20 m bathymetry exhibited higher absorption at surface waters. Absorption by picophytoplankton was strong at the bottom during the monsoon season. The phytoplankton community composition of the region and the accessory pigments' contribution to total chlorophyll-a specific absorption were ascertained by the ratio of the blue-to-red region. Between 13.33 and 1.55 was the range of the blue-to-red ratio for total phytoplankton absorption, which is expressed as aph(440)/aph(665). Small organisms like cyanobacteria and prochlorophytes dominate phytoplankton communities, as shown by the blue-to-red ratio for phytoplankton absorption, aph(440)/aph(665).

Coastal optical phenomena, predominantly attributed to Chl-a and accessory pigments, exhibit distinctive spectral variations, revealing temporal and spatial characteristics. Variations in Chl-a concentration (0.05 to 12.4 mg m^{-3}) were reflected in the phytoplankton-specific absorption coefficients, particularly aph(440) and aph(675), which ranged from 0.018 to 0.32 m^2 mg^{-1} and 0.0005 to 0.16 m^2 mg^{-1}, respectively. During monsoon months, $a_{ph}(440)$ exhibited low values (0.015 m^2 mg^{-1}) with minimal Chl-a, contrasting with higher values (up to 7.13 m^2 mg^{-1}) under elevated Chl-a conditions. The intricate relationship between aph(440) and a*ph(675) indicates non-linear absorption characteristics within algal cells, attributed to the packaging effect. This effect elucidates how higher Chl-a concentrations per cell may entail lower light absorption efficiency, owing to differences in internal pigment molecular structures among phytoplankton species. Such insights, derived from optical analyses, underscore the complexity of phytoplankton-light interactions and their implications for ecosystem dynamics and remote sensing applications.

5 Seasonal Variability Between CDOM, Phytoplankton, and Detritus

In any coastal waters, phytoplankton, CDOM, and detritus engage in a dynamic interplay, each vying for dominance in light absorption. Phytoplankton, as primary producers, harness light energy for photosynthesis, thus exerting a significant influence on the ecosystem's optical properties. CDOM, comprised of organic compounds, absorbs light primarily in the ultraviolet and blue regions of the spectrum. Its concentration fluctuates with factors like terrestrial runoff and biological activity. Detritus, composed of organic particles and debris, serves as a reservoir for nutrients and organic matter, impacting water clarity and light penetration. In the monsoon season, CDOM may exhibit a lesser contribution to total absorption compared to phytoplankton and detritus. Conversely, during post-monsoon seasons, these components may share a more equitable contribution, with detritus potentially assuming a leading role. ad(443) contributed more during the post-monsoon season, according to the ternary analysis of inherent optical properties (IOP) derived from the Ocean Color Climate Change Initiative (OC-CCI). In Kochi's coastal waters, detritus was the main light absorber, followed by phytoplankton and CDOM (Fig. 4.5).

6 Satellite Data Validation

Chl-a was operationally retrieved from CZCS, OCTS, SeaWiFS, MERIS, MODIS, and OCM2 using the six bio-optical algorithms (OC3C, OC4O, OC4, OC4E, OC3M, and OC4O2). Algorithms based on the reference band ratios 488/510/520 nm to 547/550/555/560/565 nm must be taken into account for Chl-a concentrations higher than 1.0 mg m^{-3}. The evaluation of the algorithms revealed that OC3M and OC4 performed better. In Type-I waters, all of the algorithms performed better. However, because of the substantial co-variance of Chl-a with CDOM, the performance was subpar in Type-II and Type-III waters. In the blue zone, the Chl-*a* and CDOM compete with one another for the absorption of light of similar wavelengths. Consequently, both Chl-*a* and CDOM fingerprints are present in the signal that emerges from the water column. Based on the idea that the water-leaving radiance in the blue band decreases as the Chl-*a* concentration increases, the empirical

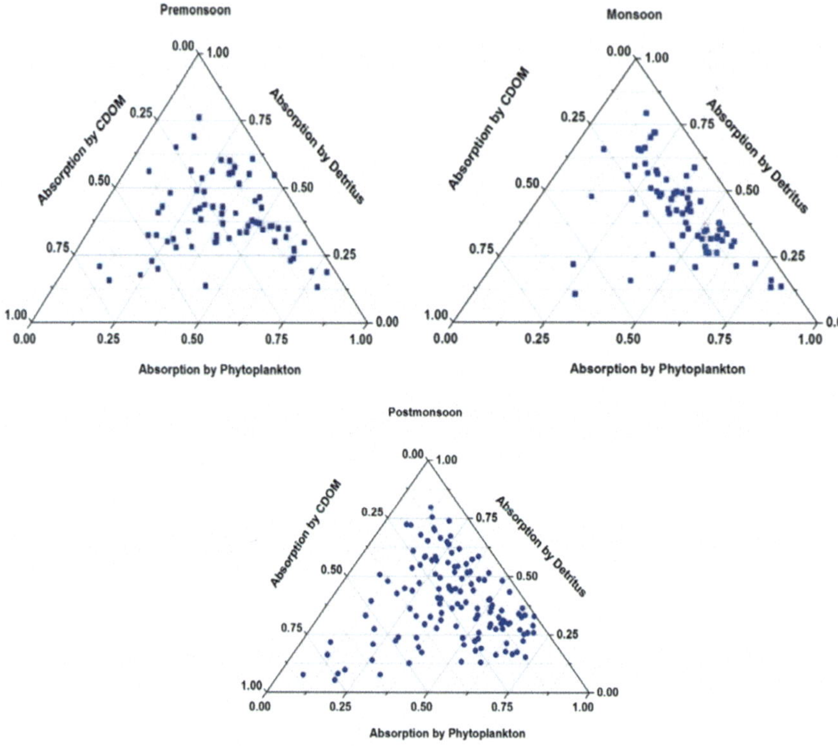

Fig. 4.5 Percentile contribution of absorption by phytoplankton, CDOM, and detritus for different seasons [pre-monsoon, monsoon, and post-monsoon]. (Source: Souda et al., 2020)

algorithms determine the ratio of the blue band to the green band. Both Chl-*a* and CDOM significantly contribute to the reduction of the water-leaving radiance in the blue band when they are the main constituents in the water column. The retrieval of Chl-*a* is weakened in such a situation by the performance of ratio-based methods.

7 Conclusion

Research on bio-optical characteristics conducted in the coastal waters near Kochi has shed light on the optical traits of these intricate coastal environments and the behavior of optically active materials. They revealed the presence of 73 genera of phytoplankton belonging to 19 orders and 41 families. Remarkably, there was a typical one-peak cycle in the annual fluctuations in phytoplankton abundance, with the pre-monsoon season having the highest levels and the monsoon season having the lowest. Furthermore, poor diversity during the monsoon season was revealed by the species diversity index. The spectral Rrs displayed three different water types

(TI, TII, and TIII), which were related to changes in OAS such as Chl-a, CDOM, and β650. The various water types exhibited distinct peak Rrs wavelengths, with TI reaching its peak at 470 nm, while TII and TIII peaked at 560 nm and 570 nm, correspondingly, reflecting changes associated with different OAS concentrations. These waters were abundant with phytoplankton pigments such as peridinin, diadinoxanthin, fucoxanthin, β-carotene, phycoerythrobilin, Chl-a, Chl-b, and Chl-c. Algorithms utilizing reference band ratios ranging from 488/510/520 nm to 547/550/555/560/565 nm were deemed effective for Chl-a concentrations exceeding 1.0 mg m^{-3}, with OC3M and OC4 algorithms exhibiting superior performance. This research greatly improves our understanding of phytoplankton behavior, their optical characteristics, and the impact on remote sensing uses in coastal environments, providing essential knowledge for both scientific research and effective management of marine resources.

Acknowledgements The authors are thankful to the Director, Central Institute of Fisheries Technology, for the support and encouragement to complete the work. The Indian National Centre for Ocean Information Services (INCOIS), a Ministry of Earth Sciences-supported institution, funded this study under the Satellite Coastal Oceanographic Research (SATCORE) program. We are also grateful to Boson and the crew of MV Bharath Darshan for the support during the cruises.

References

Aiken, J., Fishwick, J., Moore, G., & Pemberton, K. (2004). The annual cycle of phytoplankton photosynthetic quantum efficiency, pigment composition and optical properties in the western English Channel. *Journal of the Marine Biological Association of the United Kingdom, 84*(2), 301–313.

Barlow, R. G., Aiken, J., Holligan, P. M., Cummings, D. G., Maritorena, S., & Hooker, S. (2002). Phytoplankton pigment and absorption characteristics along meridional transects in the Atlantic Ocean. *Deep Sea Research Part I: Oceanographic Research Papers, 49*(4), 637–660.

Barlow, R. G., Aiken, J., Moore, G. F., Holligan, P. M., & Lavender, S. (2004). Pigment adaptations in surface phytoplankton along the eastern boundary of the Atlantic Ocean. *Marine Ecology Progress Series, 281*, 13–26.

Bhagirathan, U., Shaju, S. S., Ragesh, N., Meenakumari, B., & Ashraf, M. P. (2014). Observations on bio-optical properties of a phytoplankton bloom in coastal waters off Cochin during the onset of southwest monsoon. *Indian Journal of Geo-Marine Sciences, 43*(2), 289–296.

Ciotti, A. M., Lewis, M. R., & Cullen, J. J. (2002). Assessment of the relationships between dominant cell size in natural phytoplankton communities and the spectral shape of the absorption coefficient. *Limnology and Oceanography, 47*(2), 404–417.

Cupp, E. E. (1943). Marine plankton diatoms of the west coast of North America. *Bulletin of the Scripps Institution of Oceanography, 5*(1), 1–238.

Devassy, V. P., & Goes, J. I. (1989). Seasonal patterns of phytoplankton biomass and productivity in a tropical estuarine complex (West Coast of India). *Proceedings: Plant Science, 99*, 485–501.

Fargion, G. S., & Mueller, J. L. (2000). *Ocean Optics Protocols for SeaWiFS Validation, Revision 2* (p. 209966). NASA Technical Memorandum.

Fishwick, J. R., Aiken, J., Barlow, R., Sessions, H., Bernard, S., & Ras, J. (2006). Functional relationships and bio-optical properties derived from phytoplankton pigments, optical and photosynthetic parameters; a case study of the Benguela ecosystem. *Journal of the Marine Biological Association of the United Kingdom, 86*(6), 1267–1280.

Joshi, M., & Rao, A. D. (2012). Response of southwest monsoon winds on shelf circulation off Kerala coast, India. *Continental Shelf Research, 32*, 62–70.

Kahru, M., & Mitchell, B. G. (1998). Spectral reflectance and absorption of a massive red tide off southern California. *Journal of Geophysical Research: Oceans, 103*(C10), 21601–21609.

Kowalczuk, P., & Kaczmarek, S. (1996). Analysis of temporal and spatial variability of 'yellow substance' absorption in the southern Baltic. *Oceanologia, 38*(1), 3–32.

Madhupratap, M. S. R. S., Nair, S. S., Haridas, P., & Padmavati, G. (1990). Response of zooplankton to physical changes in the environment: Coastal upwelling along the central west coast of India. *Journal of Coastal Research, 6*, 413–426.

Minu, P., Shaju, S. S., Muhamed Ashraf, P., & Meenakumari, B. (2014). Phytoplankton community characteristics in the coastal waters of the southeastern Arabian Sea. *Acta Oceanologica Sinica, 33*, 170–179.

Minu, P., Lotliker, A. A., Shaju, S. S., Ashraf, P. M., Kumar, T. S., & Meenakumari, B. (2016). Performance of operational satellite bio-optical algorithms in different water types in the southeastern Arabian Sea. *Oceanologia, 58*(4), 317–326.

Minu, P. (2017) (awarded 2019). Phytoplankton dynamics and its influence on optical properties in the coastal waters of Southeastern Arabian Sea [Doctoral dissertation, Cochin University of Science and Technology]. Institutional Repository.

Minu, P., Souda, V. P., Baliarsingh, S. K., Dwivedi, R. M., Ali, Y., & Ashraf, P. M. (2020). Assessing temporal variation of coloured dissolved organic matter in the coastal waters of south eastern Arabian Sea. *Acta Oceanologica Sinica, 39*, 102–109.

Moore, C. M., Lucas, M. I., Sanders, R., & Davidson, R. (2005). Basin-scale variability of phytoplankton bio-optical characteristics in relation to bloom state and community structure in the Northeast Atlantic. *Deep Sea Research Part I: Oceanographic Research Papers, 52*(3), 401–419.

Morel, A. (1997). Consequences of a Synechococcus bloom upon the optical properties of oceanic (case 1) waters. *Limnology and Oceanography, 42*(8), 1746–1754.

Morel, A., & Prieur, L. (1977). Analysis of variations in ocean color 1. *Limnology and Oceanography, 22*(4), 709–722.

Shaju, S. S., Minu, P., Srikanth, A. S., Ashraf, P. M., Vijayan, A. K., & Meenakumari, B. (2015). Decomposition study of in vivo phytoplankton absorption spectra aimed at identifying the pigments and the phytoplankton group in complex case 2 coastal waters of the Arabian Sea. *Oceanological and Hydrobiological Studies, 44*(3), 282–293.

Souda, V. P., Minu, P., Lotliker, A. A., Shaju, S. S., & Ashraf, P. M. (2020). Inter-annual variability in the inherent optical properties along the southeastern Arabian Sea from 2009 to 2015. *Arabian Journal of Geosciences, 13*, 1–10.

Srinivas, K., & Kumar, P. D. (2006). Atmospheric forcing on the seasonal variability of sea level at Cochin, southwest coast of India. *Continental Shelf Research, 26*(10), 1113–1133.

Stramski, D., Bricaud, A., & Morel, A. (2001). Modeling the inherent optical properties of the ocean based on the detailed composition of the planktonic community. *Applied Optics, 40*(18), 2929–2945.

Stuart, V., Sathyendranath, S., Platt, T., Maass, H., & Irwin, B. D. (1998). Pigments and species composition of natural phytoplankton populations: Effect on the absorption spectra. *Journal of Plankton Research, 20*(2), 187–217.

Subrahmanyan, R. (1959). Studies on the phytoplankton of the west coast of India. *Proceedings: Plant Science, 50*(3), 113–187.

Tomas, C. R. (Ed.). (1997). *Identifying marine phytoplankton*. Elsevier.

Twardowski, M. S., Boss, E., Sullivan, J. M., & Donaghay, P. L. (2004). Modeling the spectral shape of absorption by chromophoric dissolved organic matter. *Marine Chemistry, 89*(1–4), 69–88.

Utermohl, H. (1958). Zur Vervollkommung der quantitativen phytoplankton-methodik. *Mitteilungen Internationale Vereinigung fur Theoretische und Angewandte Limnologie, 9*, 38.

Chapter 5
Bio-Optical Properties Associated with Coastal Waters of the North-Western Bay of Bengal

Rajdeep Roy, C. V. Jayaram, Neethu Chacko, Sanjiba Kumar Baliarsingh, P. V. Nagamani, and Aneesh A. Lotliker

Abstract The northwestern Bay of Bengal's coastal waters are heavily shaped by freshwater discharge from three major rivers: the Ganga, Brahmaputra, and Mahanadi. These inflows influence the region's bio-optical dynamics, particularly due to the presence of optically active substances like colored dissolved organic matter (CDOM), suspended sediments, and detritus—characteristic of Case-2 waters where such components behave independently from phytoplankton concentrations. Marked seasonal variations in these bio-optical properties are evident, especially during the monsoon when riverine input increases CDOM and sediment levels. Conversely, during non-monsoon periods, phytoplankton absorption becomes more prominent, closely linked with nutrient availability and light conditions, notably affecting the absorption in the blue and red wavelengths. With escalating human activity anticipated in these coastal zones, localized changes in optical properties are expected, underscoring the urgent need for improved, region-specific models to enhance the accuracy of ocean color retrievals from satellite observations.

Keywords Ocean Color · Case-2 Waters · CDOM · Phytoplankton · Hooghly River

---

R. Roy (✉) · C. V. Jayaram · N. Chacko
Regional Remote Sensing Centre—East, National Remote Sensing Centre, Indian Space Research Organization, Kolkata, India

S. K. Baliarsingh · A. A. Lotliker
Indian National Centre for Ocean Information Services, Ministry of Earth Sciences, Hyderabad, India

P. V. Nagamani
Biological Oceanography Division, Earth and Climate Science Area, National Remote Sensing Centre, Indian Space Research Organization, Hyderabad, India

© The Author(s), under exclusive license to Springer Nature Switzerland AG 2026
S. Das (ed.), *Bio-optical Properties of Indian Coastal Waters*,
https://doi.org/10.1007/978-3-032-08714-0_5

1 Coastal North-Western Bay of Bengal Vis-à-Vis Ocean Color Remote Sensing

The Bay of Bengal (BoB), a tropical, semi-enclosed marine basin, is profoundly impacted by seasonal monsoons and substantial freshwater inputs. In offshore areas, the seasonal reversal of surface currents fosters a pronounced stratification in the upper water column, which limits vertical mixing. This basin is fed by six major river systems: the Ganga, Brahmaputra, Mahanadi, Godavari, Krishna, and Kaveri, with the Ganga–Brahmaputra (G–B) system being the most dominant. These two rivers converge in Bangladesh, forming the Meghna River. The Hooghly-Sundarbans estuarine complex originates from branches of the Ganga and Brahmaputra, with the lower section comprising the Sundarbans, an extensive mangrove ecosystem. This estuary is categorized as a positive estuary and is significantly affected by anthropogenic ecological pressures (Rakshit et al., 2015). South of this system lies the Mahanadi River, the third largest river in India and the primary river in Odisha. Spanning an area of nearly 141,600 km^2 (with 65,628 km^2 in Odisha), the Mahanadi stretches 851 km and reaches a maximum discharge rate of 44,740 m^3/s (Panda et al., 2013). The river's main channel flows into the BoB at Paradip, a hub for industrial activity (Radhakrishna, 2001), while its smaller distributaries enter near Chilika Lake. Mangrove vegetation is also present within this estuarine zone (Panda et al., 2013). Urban-industrial centers such as Sambalpur, Cuttack, Bauda, Choudwar, Jagatpur, and Paradip exhibit heightened human impact (Panda et al., 2013). While the Hugli and Brahmaputra rivers are fed by glacial meltwaters, categorizing them as glacial rivers, those originating in peninsular India are primarily monsoon-driven. These monsoonal rivers receive the bulk of their freshwater inflow during the rainy season, with significantly reduced flows during the dry months (Sarma et al., 2014).

Remote sensing of ocean color relies on the interaction of light with various particles within the water column, which shape the water's inherent optical properties (IOPs) by modifying spectral signatures. Through satellite-based sensors, researchers can monitor coastal ecosystems and assess water quality indicators such as total suspended matter (TSM) and river plumes. Early observations were enabled by Landsat-1's Multispectral Scanner (MSS) (Maul & Gordon, 1975), which paved the way for the first ocean color-dedicated mission—Coastal Zone Color Scanner (CZCS) launched in 1978—allowing large-scale marine monitoring (Johnson & Harriss, 1980; Simpson & Brown, 1987). Successive missions introduced advanced sensors with better spectral and spatial resolution, including SeaWiFS (1997–2010), OCTS, OCM-1 (1999–2010), MODIS (2000 & 2002), MERIS (2002), OCM-2 (2009), VIIRS (2011), OLCI (2016), and the upcoming PACE-OCI (2024). These polar-orbiting platforms were complemented by the geostationary GOCI (2010), focused on the Korean region (Kim et al., 2014). Satellite-based mapping of suspended sediments in coastal zones has emerged as a key application, with sensors like SeaWiFS and MODIS being widely used for studying sediment distribution and transport. More recent missions, such as Landsat-8 (OLI), Landsat-9, and Sentinel-2

(MSI), have significantly enhanced high-resolution monitoring of nearshore environments. These platforms offer improved signal-to-noise ratios, enabling more accurate retrievals of water quality parameters. This advancement has supported the implementation of sophisticated atmospheric correction algorithms tailored to complex coastal waters, including C2RCC (Brockmann et al., 2016) and ACOLITE (Vanhellemont, 2019). Together, these tools form a robust foundation for assessing coastal water quality through remote sensing.

Remote sensing studies focusing on water quality in the north-western Bay of Bengal (BoB) remain relatively limited. Chacko and Jayaram (2017) investigated the long-term trends of total suspended matter (TSM) using MERIS data, revealing a clear linkage between increased river discharge during the monsoon and elevated sediment levels in the Hooghly estuary. Building on this foundation, Jayaram et al. (2021a) developed a region-specific algorithm to estimate TSM concentrations using MODIS data spanning the period from 2002 to 2016. Further contributions to this domain include assessments of seasonal variations in key water quality indicators such as chlorophyll and TSM along the northern BoB coast. These analyses employed high-resolution satellite imagery from Landsat-8 and Sentinel-2, supported by field measurements (Poddar et al., 2019; Jayaram et al., 2021b; Roy et al., 2022). This chapter focuses on examining the bio-optical characteristics of the coastal waters in the north-western BoB, particularly regions under the influence of freshwater inputs from the Ganga and Mahanadi river systems (Fig. 5.1).

1.1 *Riverine Dynamics and Circulation in North-Western Bay of Bengal*

The Hooghly River estuary exhibits pronounced spatial and temporal variability in salinity and other biogeochemical parameters, primarily driven by seasonal fluctuations in river discharge. Observations from the non-monsoon period (Fig. 5.2) highlight distinct variations in water column properties. The region experiences intense hydrodynamic activity, with nearshore current speeds frequently exceeding 100 cm/s (Sadhuram et al., 2005). The tidal regime is semi-diurnal and energetically significant, characterized by tidal amplitudes ranging from 1.8 m during neap tides to approximately 5.2 m during spring tides. Peak tidal velocities have been reported up to six knots, substantially influencing local circulation dynamics (Sadhuram et al., 2005; Mukhopadhyay et al., 2006).

Seasonal discharge in the Hooghly estuary reaches a maximum of approximately $3000 \pm 1000\ m^3\ s^{-1}$ during the southwest monsoon, declining sharply to less than $1000 \pm 80\ m^3\ s^{-1}$ in the pre-monsoon period (Sadhuram et al., 2005; Mukhopadhyay et al., 2006; Ray et al., 2015). The Ganges River, a major contributor to the estuary, transports an estimated 324×10^6 tons of sediment annually to the Bengal Fan, a phenomenon that is visibly detectable in satellite observations (Ray et al., 2015; Chacko & Jayaram, 2017).

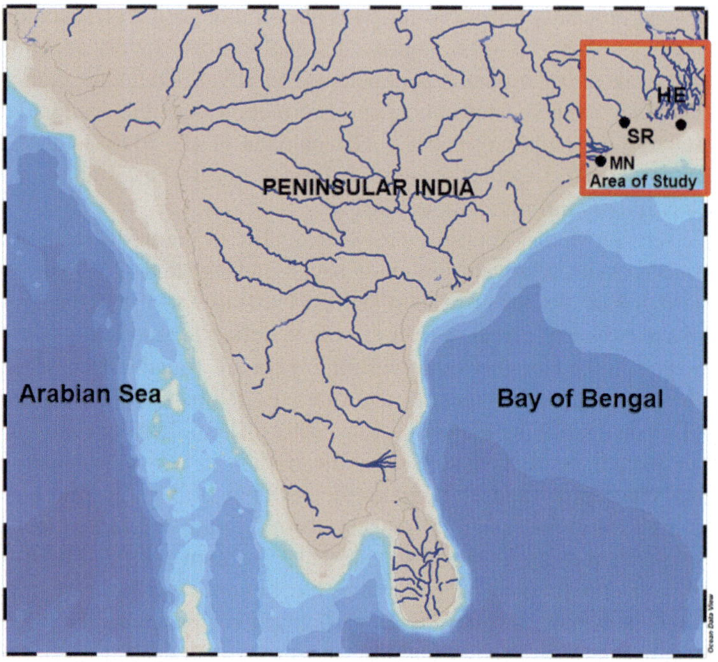

Fig. 5.1 Shows the Hooghly and Mahanadi estuary situated in the north-western Bay of Bengal. *HE* Hooghly Estuary, *SR* Subarnarekha River, *MN* Mahanadi River

In contrast, the Mahanadi River, over a 30-year period, has delivered approximately 49.0 ± 20.5 km^3 of freshwater and 17.4 ± 12.7 × 10^6 tons of sediment to the Bay of Bengal, with a mean basin-wide erosion rate of 265 ± 125 tons km^{-2} yr^{-1}. Between 2000–2001 and 2009–2010, average annual sediment and water fluxes were estimated at 12 ± 5 × 10^6 tons yr^{-1} and 49 ± 16 km^3 yr^{-1}, respectively, ranking the Mahanadi as the second-largest contributor among peninsular rivers in eastern India (Bastia & Equeenuddin, 2016).

Additionally, circulation in the north-western Bay of Bengal is modulated by the East India Coastal Current (EICC), which undergoes seasonal reversals—flowing northward from February to September with a peak in March–April, and southwestward from October to January, reaching its maximum intensity in January (Schott & McCreary, 2001).

The north-western Bay of Bengal (BoB) exhibits distinct seasonal phases, typically categorized as pre-monsoon (February–May), monsoon (June–September), and post-monsoon (October–January). Riverine inflows are a major source of nutrients to the BoB; however, a significant portion of these nutrients is retained or transformed within the estuarine regions before reaching the open ocean (Sarma et al., 2013). The combined nutrient flux from the Ganges and Brahmaputra rivers has been estimated at approximately 133 × 10^9 mol yr^{-1}, contributing about 2% of the total global riverine nutrient input to the world's oceans (Sarin et al., 1989).

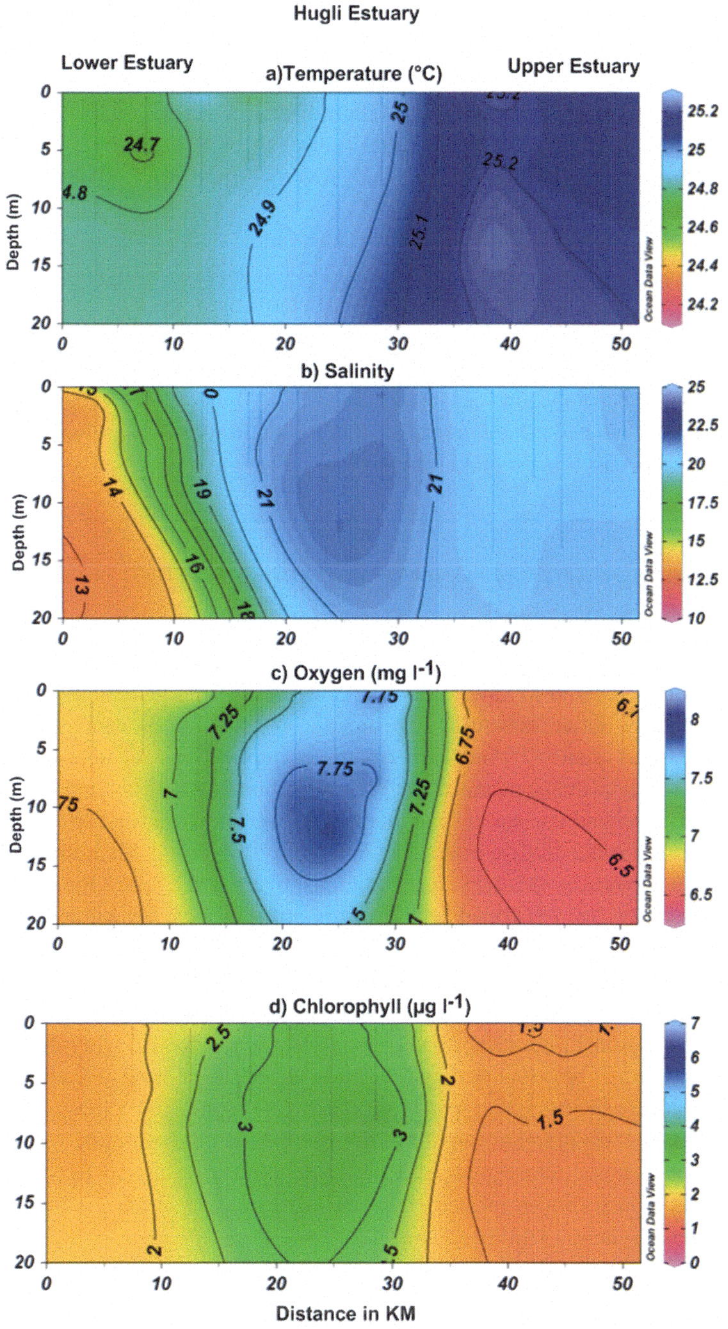

Fig. 5.2 Physical structure of Hooghly-Sundarbans estuarine region during non-monsoon season

Intensive agricultural activity along these riverbanks, supported by ample freshwater availability, further enhances nutrient loading into the system (Sarma et al., 2014). Stable nitrogen isotope ($\delta^{15}N$) analyses of particulate organic matter (POM) from Indian estuaries during periods of high discharge reveal notable spatial variability. Heavily polluted estuaries exhibit enriched $\delta^{15}N$ values exceeding 10‰, while less impacted systems show values below 6‰, indicating substantial anthropogenic influence (Sarma et al., 2014).

Furthermore, metabolic balance assessments indicate that primary production in this region is generally lower than community respiration, suggesting that the Hooghly estuary functions predominantly as a heterotrophic system (De et al., 1994).

Historical investigations indicate that diatoms represent the dominant phytoplankton group in the north-western Bay of Bengal, particularly during both pre-monsoon and post-monsoon seasons (De et al., 1994). Moreover, recurrent diatom bloom events have been documented in this region, underscoring their ecological significance in coastal phytoplankton dynamics (De et al., 1991).

1.2 Dynamics of TSM, CDOM, and Chlorophyll in North-Western Bay of Bengal

Total Suspended Matter (TSM), primarily composed of inorganic particulates, is a crucial optically active constituent in aquatic systems, significantly influencing water clarity and quality. Its presence has direct implications for light penetration, thereby affecting primary productivity and overall ecological functioning, particularly in estuarine and coastal environments. Chacko and Jayaram (2017) analyzed TSM variability in the north-western Bay of Bengal (BoB) using Medium Resolution Imaging Spectrometer (MERIS) data obtained from the GlobCOLOUR dataset (GlobCOLOUR: An EO-based service supporting global ocean carbon cycle research) (Fig. 5.3). Their findings indicate a strong correlation between TSM distribution and monsoon-driven riverine discharge in the region.

Further extending this research, Jayaram et al. (2021a) utilized MODIS-derived TSM concentrations to examine spatial variability across the Hooghly–Sundarbans estuarine system. They reported markedly elevated TSM levels, reaching up to 250 g m^{-3} during the monsoon season, in contrast to significantly lower concentrations (~50 g m^{-3}) observed during non-monsoon periods. Inter-annual variability in TSM was also noted, closely aligned with fluctuations in annual freshwater discharge. The highest concentrations were typically located near the confluence zones of the Hooghly/Ganges and Roop Narayan rivers.

The role of riverine input in modulating TSM dynamics is further corroborated by Parida et al. (2019), who reported similar patterns along the Gopalpur coast. Additionally, satellite observations show a general declining trend in TSM concentrations from the nearshore zones to offshore waters throughout the BoB coastal margin (Baliarsingh et al., 2015; Chacko & Jayaram, 2017). In the north-western

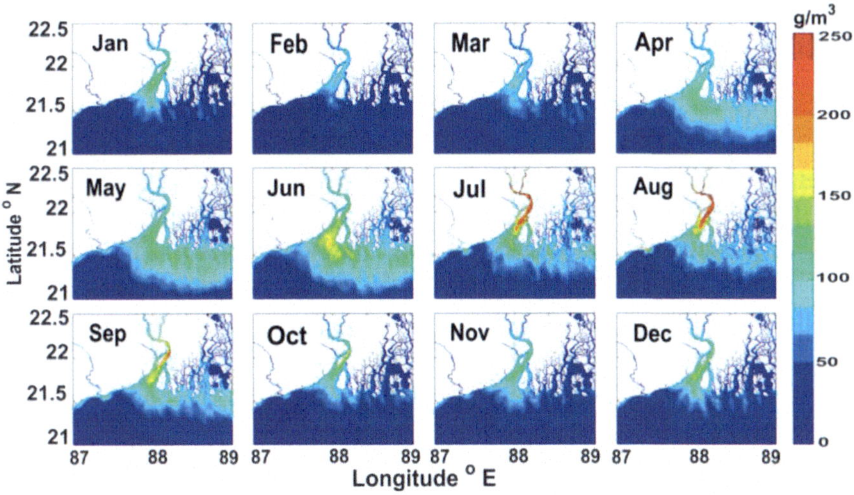

Fig. 5.3 Distribution of TSM monthly climatology in the Hooghly estuary and its adjacent riverine stretch computed from the long-term monthly data during 2003–2018

BoB, TSM distributions exhibit both seasonal and spatial heterogeneity, primarily governed by freshwater discharge, tidal mixing, and sediment resuspension in shallow shelf regions.

Colored Dissolved Organic Matter (CDOM) refers to the light-absorbing fraction of dissolved organic matter present in aquatic systems, particularly effective in absorbing light at the shorter wavelengths of the electromagnetic spectrum. CDOM primarily consists of humic substances, fulvic acids, and complex polymeric organic materials (Kowalczuk et al., 2009). In the north-western Bay of Bengal (BoB), CDOM distribution is highly variable and modulated by multiple sources, including fluvial input, phytoplankton exudates, and anthropogenic discharges.

Das et al. (2017) investigated the seasonal and spatial variability of CDOM absorption at 440 nm ($a_{CDOM}(440)$) along the continental shelf of the north-western BoB, focusing on the Hooghly estuarine region. Their findings revealed elevated $a_{CDOM}(440)$ values in nearshore waters, with a gradual decline toward offshore areas. These higher coastal concentrations were primarily attributed to terrestrial inputs, including domestic and industrial effluents delivered via riverine discharge (Bricaud et al., 1981). The study reported a distinct seasonal signal, with $a_{CDOM}(440)$ values peaking during the monsoon season, consistent with increased runoff and organic matter fluxes. Measured values of $a_{CDOM}(440)$ ranged between 0.1002 m^{-1} and 0.6631 m^{-1} over the 1-year observation period.

Supporting these findings, Parida et al. (2019) documented $a_{CDOM}(440)$ values ranging from 0.02 m^{-1} to 4.28 m^{-1} in the coastal waters off Gopalpur. The variability in CDOM absorption across different coastal regions was attributed to the distinct biogeochemical signatures of riverine inputs entering at various points along the BoB coastline. Collectively, these studies highlight the strong spatio-temporal

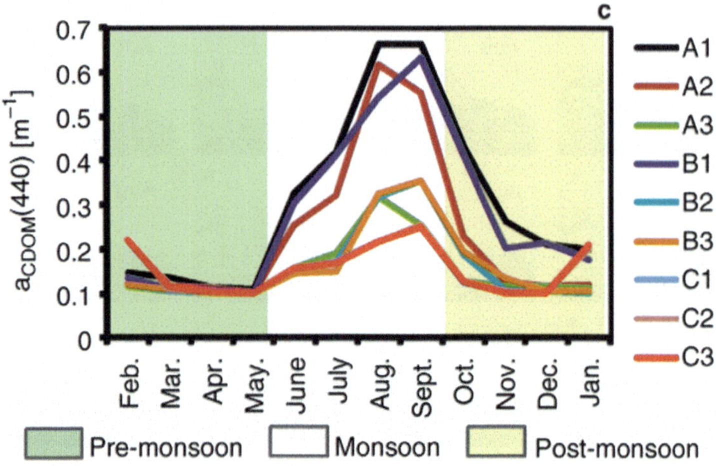

Fig. 5.4 Spatio-temporal variability of $a_{CDOM}(440)$ [m^{-1}] observed along the continental shelf of the northern coastal Bay of Bengal. (From Das et al., 2017)

variability in CDOM concentrations across the estuarine and coastal BoB, governed by freshwater discharge, biological productivity, and anthropogenic pressures (Fig. 5.4).

Chlorophyll, the primary photosynthetic pigment in phytoplankton, serves as a key indicator of phytoplankton biomass and is widely utilized as a proxy for estimating primary productivity in marine ecosystems. In the coastal and estuarine waters of the Bay of Bengal (BoB), chlorophyll concentration exhibits significant spatial and temporal variability, predominantly influenced by riverine nutrient input, seasonal monsoonal forcing, tidal mixing, and light availability (Lotliker et al., 2020).

Using MODIS Level-4 chlorophyll products, Singh and Chaturvedi (2010) documented elevated chlorophyll concentrations during the winter months, particularly in the deltaic zones influenced by the Ganga and Mahanadi River systems. This seasonal enhancement is likely driven by improved water clarity, reduced stratification, and increased nutrient availability associated with post-monsoonal mixing. In contrast, the southern coastal BoB exhibited relatively subdued seasonal variability in chlorophyll concentrations, possibly due to lower riverine influence and reduced nutrient fluxes.

However, a distinct seasonal signal is evident along the northern coastal BoB, particularly in the ecologically sensitive Gopalpur region. Parida et al. (2019) reported pronounced seasonal variability in chlorophyll levels in these waters, suggesting a strong coupling with monsoonal discharge and associated nutrient dynamics. Additionally, Baliarsingh et al. (2015) observed a bimodal seasonal pattern in surface chlorophyll concentrations off Rushikulya, with two distinct peaks occurring in March and September, corresponding to the pre-monsoon and post-monsoon transition phases, respectively (Fig. 5.5). These findings underscore the complex

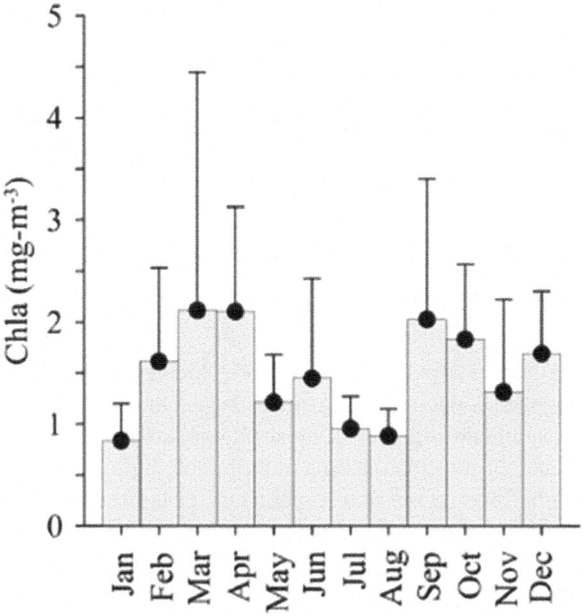

Fig. 5.5 Seasonal variability of chlorophyll concentration observed at Rushikulya. (Adapted from Baliarsingh et al., 2015)

interplay between hydrodynamic, biogeochemical, and climatic drivers regulating phytoplankton dynamics in the coastal BoB.

Lotliker et al. (2020) reported a distinct pre-monsoon peak in surface chlorophyll concentrations during March–April at a coastal site in the north-western Bay of Bengal (BoB), indicative of enhanced phytoplankton productivity during this period. Chlorophyll concentrations in this region exhibit substantial variability, ranging from 0.1 to 10 mg m^{-3}, largely influenced by seasonal riverine input, nutrient fluxes, and hydrodynamic processes. Spatially, chlorophyll levels generally decline with increasing distance from the shoreline; Dey and Singh (2003) observed a significant decrease in concentration beyond 11–12 km offshore, suggesting that phytoplankton biomass is predominantly concentrated in nearshore zones where nutrient availability is higher.

In addition to seasonal and spatial drivers, episodic events such as tropical cyclones also play a critical role in modulating surface chlorophyll dynamics. Several studies have demonstrated that cyclone-induced mixing and upwelling can significantly enhance nutrient availability in the euphotic zone, thereby stimulating phytoplankton growth. Enhanced chlorophyll concentrations following cyclone activity have been documented in the BoB by Kundu et al. (2001) and Vinayachandran and Mathew (2003), emphasizing the influence of atmospheric forcing on biological productivity.

Furthermore, Baliarsingh et al. (2013) analyzed chlorophyll distribution along various coastal transects in the BoB and reported concentrations ranging from 1.5 to 8.5 mg m^{-3}, with an average of approximately 4 mg m^{-3}. These findings underscore the strong spatial heterogeneity in chlorophyll distribution, shaped by localized environmental conditions, hydrological regimes, and episodic weather events.

1.3 Phytoplankton Absorption and Remote Sensing Reflectance

Phytoplankton utilize solar radiation absorbed by their pigments to drive photosynthesis, facilitating the production of organic carbon compounds essential for marine food webs. The dominant pigment in most phytoplankton species, chlorophyll-a, exhibits two primary absorption maxima in the visible light spectrum: a major peak near 440 nm in the blue region and a secondary peak around 675 nm in the red region (Fig. 5.6). The spectral absorption profile is further influenced by accessory pigments—such as carotenoids and phycobiliproteins—which vary across taxa and broaden the blue absorption peak or introduce additional peaks depending on pigment composition (Subramaniam & Carreto, 2006). These species-specific absorption characteristics form the basis of bio-optical techniques used for in situ identification and are integral to the calibration of remote sensing algorithms.

Phytoplankton absorption significantly modulates the underwater light field, affecting the quantity and quality of light available at depth (Gordon et al., 1988; Platt & Sathyendranath, 1988). Consequently, many physiological and ecological models of primary production depend on accurate representation of the in vivo phytoplankton absorption coefficient, aph(λ), which describes the inherent ability of phytoplankton to absorb light at a given wavelength λ (Cullen, 1990). This parameter also plays a central role in determining water-leaving radiance and remote

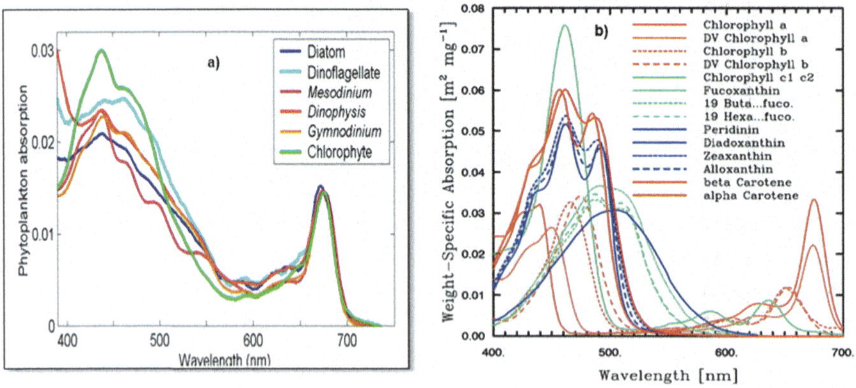

Fig. 5.6 Chlorophyll normalized phytoplankton absorption based on Bricaud et al. (1995) and Ciotti et al. (2002)

sensing reflectance, making it critical for satellite-based estimations of chlorophyll-a and, by extension, phytoplankton biomass (Gordon et al., 1983). Several bio-optical models have been developed to estimate oceanic primary production from satellite data, and these are strongly dependent on the absorption characteristics of phytoplankton (Platt & Sathyendranath, 1988; Sathyendranath & Morel, 1983; Sathyendranath & Platt, 1989).

An important derived metric is the specific absorption coefficient, aph*(λ), defined as aph(λ) normalized by the chlorophyll-a concentration. Initially treated as a constant across wavelengths in early modeling efforts (Morel, 1991), aph*(λ) is now recognized to be highly variable, influenced by species composition, pigment packaging effects, and physiological state (Bidigare et al., 1990). Observations show that aph*(440) tends to be lower in stratified open-ocean waters compared to well-mixed coastal zones—largely due to differences in phytoplankton community structure. Notably, in coastal environments, pronounced spectral variability is often observed in the 510–570 nm range. An increase in specific absorption in this band has been attributed to elevated carotenoid content relative to chlorophyll-a. Additionally, a strong correlation between relative absorption at 550 nm and the fucoxanthin-to-chlorophyll-a ratio highlights the influence of diatom dominance in these waters.

In addition to absorption, the scattering properties of phytoplankton are critical for interpreting ocean color remote sensing data, as they directly influence the remote sensing reflectance through the ratio of backscattering to absorption. The total scattering and backscattering coefficients of phytoplankton, along with the volume scattering function, are essential optical parameters that can be derived either through theoretical modeling—such as Mie theory—or via in situ measurements (Bricaud et al., 1983; Volten et al., 1998). These scattering characteristics are primarily determined by the physical attributes of phytoplankton cells, including their size distribution, morphology, and the complex refractive index of intracellular components (Sullivan & Twardowski, 2009).

Compared to other particulate constituents in the ocean, phytoplankton generally exhibit relatively low scattering efficiency. This is attributed to their high internal water content and strong pigment absorption, both of which reduce their overall scattering cross-section. However, a notable exception is the coccolithophores, a group of phytoplankton characterized by the production of calcite platelets known as coccoliths. These mineralized structures significantly enhance light scattering, making coccolithophore blooms highly reflective and readily detectable via satellite sensors (Balch et al., 1996). As such, understanding the scattering behavior of different phytoplankton taxa is essential for accurate remote sensing retrievals and for distinguishing between biological and non-biological components in ocean color imagery.

Field-based optical remote sensing relies on the measurement of upwelling radiance (Lw) and downwelling irradiance (Ed) to derive a fundamental optical parameter known as remote sensing reflectance (Rrs, units: sr^{-1}). Rrs is defined as the ratio of water-leaving radiance (Lw, measured in $\mu W\ cm^{-2}\ sr^{-1}\ nm^{-1}$) to the downwelling irradiance just above the sea surface (Es, measured in $\mu W\ cm^{-2}\ nm^{-1}$). This parameter serves as a key input for estimating a range of biogeochemical and inherent optical properties of natural waters, including chlorophyll-a concentration (O'Reilly et al., 1998), colored dissolved organic matter (CDOM) absorption (Lee et al., 2004), and particulate backscattering coefficients (Mannino et al., 2008; Wei & Lee, 2015). Given its central role in bio-optical modeling and satellite algorithm development, the accurate derivation of Rrs is crucial.

Remote sensing reflectance is not measured directly; rather, it is calculated from independently acquired field measurements of Lw and Es. One common methodology involves the in-water approach, wherein downwelling irradiance (Ed) and upwelling radiance (Lu) are recorded below the surface. The upwelling radiance is then corrected for water–air interface effects and propagated to just above the water surface to obtain Lw, from which Rrs is derived.

In the north-western Bay of Bengal (BoB), in situ studies have shown that the spectral shape of Rrs is sensitive to phytoplankton composition, particularly the prevalence of diatoms in surface waters. These variations in Rrs are reflective of vertical changes in the phytoplankton community structure and associated pigment composition (Fig. 5.7). Spectrally, absorption by dissolved and particulate matter is most prominent in the blue region of the visible spectrum (around 400–500 nm), while absorption in the red region (around 675 nm) is primarily due to water itself and the red absorption peak of chlorophyll-a.

Empirical relationships established between in situ remote sensing reflectance (Rrs) and chlorophyll-a concentrations are fundamental for the formulation of region-specific bio-optical algorithms. These algorithms enable improved estimates of phytoplankton biomass and other optical properties from satellite observations (O'Reilly et al., 1998). Moreover, rigorous validation of in situ Rrs against satellite-derived reflectance data is a critical step toward the accurate retrieval of geophysical parameters from spaceborne sensors. Such validation ensures the reliability and applicability of satellite-based ocean color products for biogeochemical assessments. However, comprehensive investigations addressing these aspects remain limited for the north-western Bay of Bengal, highlighting a significant gap in current research. This presents an important opportunity to enhance our understanding of coastal bio-optical variability and improve regional remote sensing capabilities.

2 Conclusion

The synthesis of available literature on the bio-optical characteristics of the north-western Bay of Bengal (BoB) indicates that, although significant progress has been made over the past decade in estimating suspended sediment and chlorophyll concentrations via remote sensing, comprehensive radiometric measurements and phytoplankton absorption data remain largely underrepresented. These foundational datasets are critical for the development, calibration, and validation of regionally tuned bio-optical algorithms, particularly in optically complex (Case-2) waters. The bio-optical variability in this region is strongly regulated by seasonal riverine

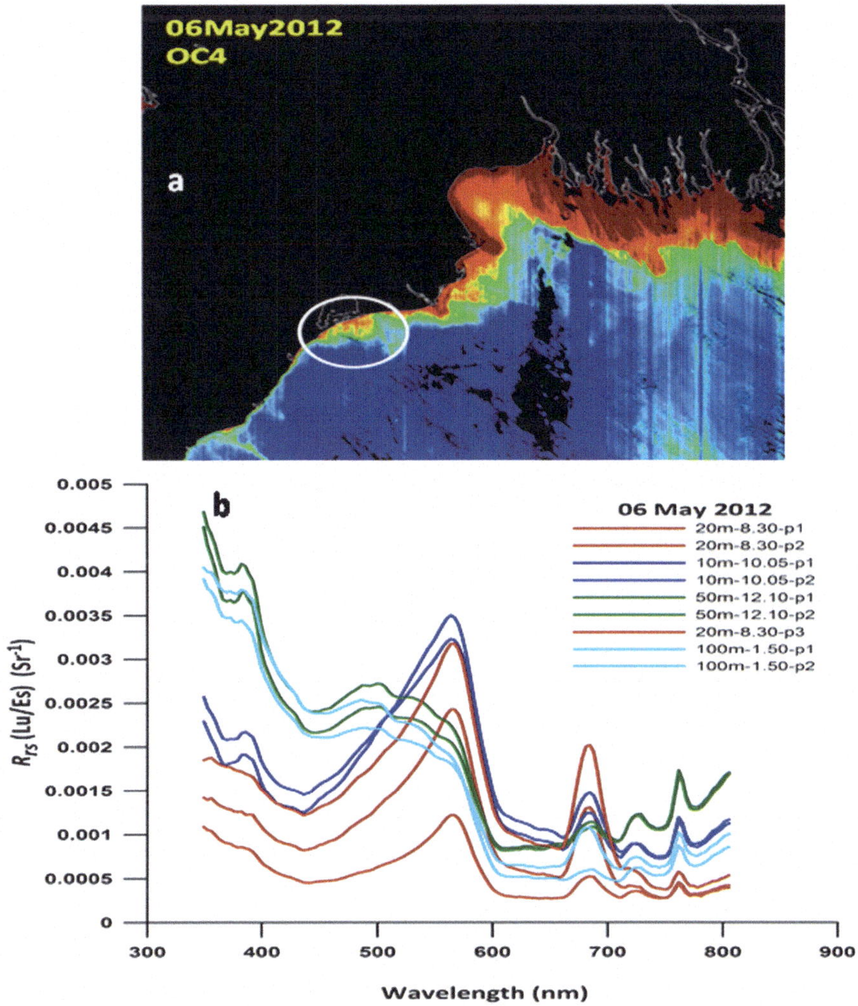

Fig. 5.7 Shows the in situ derived R_{rs} associated with a diatom bloom sampled in 2012; (**a**) Ocean color monitor (onboard Oceansat 2) image of the phytoplankton bloom (**b**) Radiometric measurements and derived R_{rs} for the same

discharge, which modulates the underwater light climate and, consequently, phytoplankton community structure and productivity. Given this complexity, periodic validation of satellite-derived chlorophyll estimates is necessary to account for sediment-induced optical interference. With the deployment of next-generation satellite missions such as OceanSat-3 and NASA's PACE (Plankton, Aerosol, Cloud, ocean Ecosystem), there is a promising outlook for advancing retrieval techniques and improving the accuracy of satellite ocean color products in coastal and estuarine systems. Continued in situ observations and algorithm refinement will be vital for enhancing our understanding of biogeochemical processes in the north-western BoB and for enabling robust regional applications of ocean color remote sensing.

Acknowledgements Rajdeep Roy, C. V. Jayaram, Neethu Chacko acknowledges, Director, National Remote Sensing Centre (NRSC) Hyderabad, Chief General Manager and Deputy General Manager Regional Remote Sensing Centre, East, NRSC, ISRO for their constant encouragement. Rajdeep Roy, Aneesh A. Lotliker, and Sanjiba Kumar Baliarsingh, are thankful to Director, Indian National Centre for Ocean Information Services (INCOIS), Ministry of Earth Sciences, Hyderabad, India, for their constant encouragement. This is INCOIS contribution no 569.

References

Balch, W. M., Kilpatrick, K. A., Holligan, P., Harbour, D., & Fernandez, E. (1996). The 1991 coccolithophore bloom in the Central North Atlantic. 2. Relating optics to coccolith concentration. *Limnology and Oceanography, 41*(8), 1684–1696.

Baliarsingh, S. K., Srichandan, S., Naik, S., Sahu, K. C., Lotliker, A. A., & Kumar, T. S. (2013). Distribution of hydro-biological parameters in coastal waters off Rushikulya estuary, east coast of India: A premonsoon case study. *Pakistan Journal of Biological Sciences, 16*(16), 779–787.

Baliarsingh, S. K., Lotliker, A. A., Sahu, K. C., & Kumar, T. S. (2015). Spatio temporal distribution of chlorophyll-a in relation to physicochemical parameters in coastal waters of the northwestern Bay of Bengal. *Environmental Monitoring and Assessment, 187*(7), 481.

Bastia, F., & Equeenuddin, S. M. (2016). Spatio-temporal variation of water flow and sediment discharge in the Mahanadi River, India. *Global and Planetary Change, 144*, 51–66.

Bidigare, R. R., Ondrusek, M. E., Morrow, J. H., & Kiefer, D. A. (1990). In-vivo absorption properties of algal pigments. In *Ocean optics X* (Vol. 1302, pp. 290–302). SPIE.

Bricaud, A., Morel, A., & Prieur, L. (1981). Absorption by dissolved organic matter of the sea (yellow substance) in the UV and visible domains. *Limnology and Oceanography, 26*(1), 43–53. https://doi.org/10.4319/lo.1981.26.1.0043

Bricaud, A., Morel, A., & Prieur, L. (1983). Optical efficiency factors of some phytoplankters 1. *Limnology and Oceanography, 28*(5), 816–832.

Bricaud, Annick, Marcel Babin, André Morel, and Hervé Claustre. (1995). Variability in the chlorophyll-specific absorption coefficients of natural phytoplankton: Analysis and parameterization. *Journal of Geophysical Research: Oceans 100*, C7: 13321–13332.

Brockmann, C., Doerffer, R., Peters, M., Kerstin, S., Embacher, S., & Ruescas, A. (2016). Evolution of the C2RCC neural network for Sentinel 2 and 3 for the retrieval of ocean colour products in normal and extreme optically complex waters. In *The Proceedings of Living Planet Symposium 2016, Prague, Czech Republic* (Vol. 740, p. 54).

Chacko, N., & Jayaram, C. (2017). Variability of total suspended matter in the northern coastal Bay of Bengal as observed from satellite data. *Journal of the Indian Society of Remote Sensing, 45*, 1077–1083. https://doi.org/10.1007/s12524-016-0650-x

Ciotti, Aurea M., Marlon R. Lewis, and John J. Cullen. (2002). Assessment of the relationships between dominant cell size in natural phytoplankton communities and the spectral shape of the absorption coefficient. *Limnology and Oceanography 47*(2), 404–417.

Cullen, J. J. (1990). On models of growth and photosynthesis in phytoplankton. *Deep Sea Research Part A. Oceanographic Research Papers, 37*(4), 667–683.

Das, S., Das, I., Giri, S., Chanda, A., Maity, S., Lotliker, A. A., Srinivasa Kumar, T., Akhand, A., & Hazra, S. (2017). Chromophoric dissolved organic matter (CDOM) variability over the continental shelf of the northern bay of Bengal. *Oceanologia, 59*(3), 271–282.

De, T. K., Ghosh, S. K., Jana, T. K., & Choudhury, A. (1991). Phytoplankton bloom in the Hooghly estuary. *Indian Journal of Marine Sciences, 20*, 134–137.

De, T. K., Choudhury, A., & Jana, T. K. (1994). Phytoplankton community organization and species diversity in the Hugli estuary, northeast coast of India. *Indian Journal of Marine Sciences, 23*, 152–156.

Dey, S., & Singh, R. P. (2003). Comparison of chlorophyll distributions in the northeastern Arabian Sea and southern bay of Bengal using IRS-P4 ocean color monitor data. *Remote Sensing of Environment, 85*, 424–428.

Gordon, H. R., Morel, A. Y., Gordon, H. R., & Morel, A. Y. (1983). Physics of ocean color remote sensing. In *Remote assessment of ocean color for interpretation of satellite visible imagery: A review* (pp. 3–23).

Gordon, H. R., Brown, O. B., Evans, R. H., Brown, J. W., Smith, R. C., Baker, K. S., & Clark, D. K. (1988). A semianalytic radiance model of ocean color. *Journal of Geophysical Research: Atmospheres, 93*(D9), 10909–10924.

Jayaram, C., Patidar, G., Swain, D., Chowdary, V. M., & Bandyopadhyay, S. (2021a). Total suspended matter distribution in the Hooghly river estuary and the Sundarbans: A remote sensing approach. *IEEE Journal of Selected Topics in Applied Earth Observations and Remote Sensing, 14*, 9064–9070. https://doi.org/10.1109/JSTARS.2021.3076104

Jayaram, C., Roy, R., Chacko, N., Swain, D., Ramunaidu, P., Bandyopadhyay, S., Choudhury, S. B., & Dutta, D. (2021b). Anomalous reduction of the total suspended matter during COVID-19 lockdown in the Hooghly estuarine system. *Frontiers in Marine Science, 8*, 633493. https://doi.org/10.3389/fmars.2021.633493

Johnson, R. W., & Harriss, R. C. (1980). Remote sensing for water quality and biological measurements in coastal waters. *Photogrammetric Engineering and Remote Sensing, 46*, 77–85.

Kim, Y. H., Im, J., Ha, H. K., Choi, J. K., & Ha, S. (2014). Machine learning approaches to coastal water quality monitoring using GOCI satellite data. *GIScience & Remote Sensing, 51*, 158–174. https://doi.org/10.1080/15481603.2014.900983

Kowalczuk, P., Durako, M. J., Young, H., Kahn, A. E., Cooper, W. J., & Gonsior, M. (2009). Characterization of dissolved organic matter fluorescence in the South Atlantic bight with use of PARAFAC model: Interannual variability. *Marine Chemistry, 113*(3–4), 182–196.

Kundu, S. N., Sahoo, A. K., Mohapatra, S., & Singh, R. P. (2001). Change analysis using IRS-P4 OCM data after the Orissa super cyclone. *International Journal of Remote Sensing, 22*(7), 1383–1389. https://doi.org/10.1080/01431160119932

Lee, Z. P., Carder, K. L., & Du, K. P. (2004). Effects of molecular and particle scatterings on the model parameter for remote-sensing reflectance. *Applied Optics, 43*(25), 4957–4964.

Lotliker, A. A., Baliarsingh, S. K., Sahu, K. C., et al. (2020). Long-term chlorophyll-*a* dynamics in tropical coastal waters of the western Bay of Bengal. *Environmental Science and Pollution Research, 27*, 6411–6419. https://doi.org/10.1007/s11356-019-07403-0

Mannino, A., Russ, M. E., & Hooker, S. B. (2008). Algorithm development and validation for satellite-derived distributions of DOC and CDOM in the US middle Atlantic bight. *Journal of Geophysical Research: Oceans, 113*(C7).

Maul, G. A., & Gordon, H. R. (1975). On the use fo earth resources technology satellite (Landsat-1) in optical oceanography. *Remote Sensing of Environment, 4*, 95–128.

Morel, A. (1991). Light and marine photosynthesis: A spectral model with geochemical and climatological implications. *Progress in Oceanography, 26*(3), 263–306.

Mukhopadhyay, S. K., Biswas, H. D. T. K., De, T. K., & Jana, T. K. (2006). Fluxes of nutrients from the tropical river Hooghly at the land–ocean boundary of Sundarbans, NE coast of Bay of Bengal, India. *Journal of Marine Systems, 62*(1–2), 9–21.

O'Reilly, J. E., Maritorena, S., Mitchell, B. G., Siegel, D. a., Carder, K. L., Garver, S. a., Kahru, M., & McClain, C. (1998). Ocean color chlorophyll algorithms for SeaWiFS. *Journal of Geophysical Research: Oceans, 103*(C11), 24937–24953.

Panda, D. K., Kumar, A., Ghosh, S., & Mohanty, R. K. (2013). Streamflow trends in the Mahanadi River basin (India): Linkages to tropical climate variability. *Journal of Hydrology, 495*, 135–149.

Parida, C., Baliarsingh, S. K., Lotliker, A. A., Dash, M., Srichandan, S., & Sahu, K. C. (2019). Seasonal variation in optically active substances at a coastal site along western Bay of Bengal. *SN Applied Sciences, 1*, 1210. https://doi.org/10.1007/s42452-019-1257-y

Platt, T., & Sathyendranath, S. (1988). Oceanic primary production: Estimation by remote sensing at local and regional scales. *Science, 241*(4873), 1613–1620.

Poddar, S., Chacko, N., & Swain, D. (2019). Estimation of chlorophyll-a in northern coastal bay of Bengal using Landsat-8 OLI and Sentinel-2 MSI sensors. *Frontiers in Marine Science, 6*, 598. https://doi.org/10.3389/fmars.2019.00598

Radhakrishna, I. (2001). Saline fresh water interface structure in Mahanadi delta region, Orissa, India. *Environmental Geology, 40*(3), 369–380.

Rakshit, D., Sarkar, S. K., Bhattacharya, B. D., Jonathan, M. P., Biswas, J. K., Mondal, P., & Mitra, S. (2015). Human-induced ecological changes in western part of Indian Sundarban megadelta: A threat to ecosystem stability. *Marine Pollution Bulletin, 99*(1–2), 186–194.

Ray, R., Rixen, T., Baum, A., Malik, A., Gleixner, G., & Jana, T. K. (2015). Distribution, sources and biogeochemistry of organic matter in a mangrove dominated estuarine system (Indian Sundarbans) during the pre-monsoon. *Estuarine, Coastal and Shelf Science, 167*, 404–413.

Roy, R., Jayaram, C., Naidu, P. R., Chandrasekar, K., Mishra, S., Mohammed, S. K., & Chacko, N. (2022). Northeast monsoon distribution of biogeochemical properties and phytoplankton pigments from Hooghly-Sundarbans estuarine region, India. *Regional Studies in Marine Science, 56*, 102703. https://doi.org/10.1016/j.rsma.2022.102703

Sadhuram, Y., Sarma, V. V., Ramana Murthy, T. V., & Prabhakara Rao, B. (2005). Seasonal variability of physico-chemical characteristics of the Haldia channel of Hooghly estuary, India. *Journal of Earth System Science, 114*, 37–49.

Sarin, M. M., Krishnaswami, S., Kadir Dilli, Somayajulu, B. L. K., & Moore, W. S. (1989). Major ion chemistry of the ganga-Brahmaputra river system: Weathering processes and fluxes to the Bay of Bengal. *Geochimica et Cosmochimica Acta, 53*(5), 997–1009.

Sarma, V. V. S. S., Krishna, M. S., Viswanadham, R., Rao, G. D., Rao, V. D., Sridevi, B., Kumar, B. S. K., et al. (2013). Intensified oxygen minimum zone on the western shelf of bay of Bengal during summer monsoon: Influence of river discharge. *Journal of Oceanography, 69*, 45–55.

Sarma, V. V. S. S., Krishna, M. S., Prasad, V. R., Kumar, B. S. K., Naidu, S. A., Rao, G. D., Viswanadham, R., Sridevi, T., Kumar, P. P., & Reddy, N. P. C. (2014). Distribution and sources of particulate organic matter in the Indian monsoonal estuaries during monsoon. *Journal of Geophysical Research: Biogeosciences, 119*(11), 2095–2111.

Sathyendranath, S., & Morel, A. (1983). Light emerging from the sea—Interpretation and uses in remote sensing. In *Remote sensing applications in marine science and technology* (pp. 323–357). Springer.

Sathyendranath, S., & Platt, T. (1989). Remote sensing of ocean chlorophyll: Consequence of nonuniform pigment profile. *Applied Optics, 28*(3), 490–495.

Schott, F. A., & McCreary Jr, J. P. (2001). The monsoon circulation of the Indian Ocean. *Progress in Oceanography, 51*(1), 1–123.

Simpson, J. H., & Brown, J. (1987). The interpretation of visible band imagery of turbid shallow seas in terms of the distribution of suspended particulates. *Continental Shelf Research, 7*, 1307–1313.

Singh, R. P., & Chaturvedi, P. (2010). Comparison of chlorophyll concentration in the Bay of Bengal and the Arabian Sea using IRS-P4 OCM and MODIS aqua. *Indian Journal of Marine Sciences, 39*(3), 334–340.

Subramaniam, A., & Carreto, J. I. (2006). Annual variations in bio-optical properties at the 'Estación Permanente de Estudios Ambientales (EPEA)' coastal station. *Continental Shelf Research, 26*, 1093–1112.

Sullivan, J. M., & Twardowski, M. S. (2009). Angular shape of the oceanic particulate volume scattering function in the backward direction. *Applied Optics, 48*(35), 6811–6819.

Vanhellemont, Q. (2019). Adaptation of the dark spectrum fitting atmospheric correction for aquatic applications of the Landsat and Sentinel-2 archives. *Remote Sensing of Environment, 225*, 175–192.

Vinayachandran, P. N., and Simi Mathew. (2003). Phytoplankton bloom in the Bay of Bengal during the northeast monsoon and its intensification by cyclones. *Geophysical Research Letters, 30*, 11.

Volten, H., De Haan, J. F., Hovenier, J. W., Schreurs, R., Vassen, W., Dekker, A. G., Hoogenboom, H. J., Charlton, F., & Wouts, R. (1998). Laboratory measurements of angular distributions of light scattered by phytoplankton and silt. *Limnology and Oceanography, 43*(6), 1180–1197.

Wei, J., & Lee, Z. (2015). Retrieval of phytoplankton and colored detrital matter absorption coefficients with remote sensing reflectance in an ultraviolet band. *Applied Optics, 54*(4), 636–649.

Chapter 6
Phytoplankton Size Structure and Functional Types in Indian Coastal Waters: A Brief Overview

Manoj Kumar Das and Tamoghna Acharyya

Abstract Phytoplankton communities play a crucial role in marine ecosystems, serving as primary producers and key components of marine food webs. Understanding the size structure and functional types of phytoplankton in Indian coastal waters is essential for assessing ecosystem health and productivity. This brief overview aims to summarize the current knowledge on phytoplankton size structure and functional types in these waters.

The size structure of phytoplankton in Indian coastal waters varies widely, ranging from picoplankton (<2 μm) to microplankton (>20 μm). Factors controlling phytoplankton size include nutrient availability, light intensity, temperature, and grazing pressure. Nutrient availability, particularly nitrogen and phosphorus, often limits phytoplankton growth and influences size distribution.

Functional types of phytoplankton refer to different groups based on their ecological roles and physiological characteristics. Common functional types include diatoms, dinoflagellates, coccolithophores, and cyanobacteria. Diatoms are prevalent in nutrient-rich coastal waters, contributing significantly to primary production and forming the base of the food web. Dinoflagellates are known for their ability to form harmful algal blooms under certain conditions, impacting ecosystem dynamics and human health. Coccolithophores play a crucial role in carbon cycling and ocean acidification, while cyanobacteria are important nitrogen fixers, particularly in nutrient-poor waters.

Various environmental factors, including temperature, salinity, nutrient availability, and grazing pressure, regulate the distribution and abundance of these functional groups. Anthropogenic activities, such as coastal development and pollution, can also influence phytoplankton communities and functional types, leading to shifts in ecosystem structure and function. Understanding the size structure and functional types of phytoplankton in Indian coastal waters is essential for ecosystem management and conservation efforts.

M. K. Das · T. Acharyya (✉)
Department of Marine Sciences, Berhampur University, Berhampur, Odisha, India
e-mail: ta.ms@buodisha.edu.in

© The Author(s), under exclusive license to Springer Nature Switzerland AG 2026

S. Das (ed.), *Bio-optical Properties of Indian Coastal Waters*,
https://doi.org/10.1007/978-3-032-08714-0_6

Keywords Phytoplankton · Marine ecosystem · Functional types · Size structure · Indian Coastal Waters

1 Introduction

Phytoplankton, the microscopic algae inhabiting marine ecosystems, play a pivotal role as primary producers and act as foundation of marine food webs. These photosynthetic organisms are essential for sustaining marine life, contributing to the global carbon cycle (Behrenfeld et al., 2005; Giering et al., 2014), and supporting fisheries and biodiversity (Friedland et al., 2012; Falkowski, 2012). Understanding the size structure and functional types of phytoplankton in Indian coastal waters is important for assessing ecosystem health and productivity, as these characteristics are influenced by various environmental factors and human activities.

Phytoplankton size structure, which ranges from picoplankton (< 2 μm) to microplankton (> 20 μm), is a fundamental aspect that determines their ecological roles and interactions within the marine environment (Chisholm, 1992). The distribution of these size classes is regulated by nutrient availability, light intensity, temperature, and grazing pressure, which together shape the dynamics of phytoplankton communities (Finkel et al., 2010). For instance, nutrient-rich conditions typically favor larger phytoplankton, such as diatoms, which can dominate primary production and form the base of the marine food web (Boyd et al., 2010).

Functional types of phytoplankton, categorized based on their ecological roles and physiological characteristics, include diatoms, dinoflagellates, coccolithophores, and cyanobacteria. Diatoms are particularly prevalent in nutrient-rich coastal waters, contributing significantly to primary production and biogeochemical cycling (Sarthou et al., 2005). Dinoflagellates, known for their ability to form harmful algal blooms, can drastically alter ecosystem dynamics and pose risks to human health (Anderson et al., 2012). Coccolithophores play a critical role in carbon cycling and ocean acidification processes, while cyanobacteria are important nitrogen fixers, especially in nutrient-poor environments (Paerl, 2008).

The distribution and abundance of these functional groups are regulated by environmental factors such as temperature, salinity, nutrient availability, and grazing pressure (Tables 6.1 and 6.2). Additionally, anthropogenic activities, including coastal development and pollution, significantly influence phytoplankton communities, leading to shifts in ecosystem structure and function (Hallegraeff, 1993). The ongoing environmental changes and human impacts in coastal ecosystems necessitate a thorough understanding of phytoplankton dynamics to inform effective ecosystem management and conservation strategies.

Table 6.1 Factors controlling phytoplankton size structure

Factor	Description	Reference
Nutrient availability	Key factor influencing phytoplankton size structure. Larger phytoplankton dominate in nutrient-rich environments	Finkel et al. (2010)
Light intensity	Affects growth and size distribution. Smaller phytoplankton adapt better to low light	Irwin et al. (2015)
Temperature	Determines metabolic rates and growth. Smaller phytoplankton thrive in warmer waters	Marañón (2015)
Grazing pressure	Major control on size structure. Grazing pressure can remove larger phytoplankton	Calbet and Landry (2004)

Table 6.2 Major functional types of phytoplankton

Functional type	Ecological role	Reference
Diatoms	Primary production, biogeochemical cycling	Sarthou et al. (2005)
Dinoflagellates	Harmful algal blooms, ecosystem dynamics	Anderson et al. (2012)
Coccolithophores	Carbon cycling, ocean acidification	Rost and Riebesell (2004)
Cyanobacteria	Nitrogen fixation, primary production in nutrient-poor environments	Paerl (2008)

1.1 Classification of Phytoplankton by Size

Phytoplankton are typically classified into three main size categories: picoplankton, nanoplankton, and microplankton. Picoplankton, which are less than 2 micrometers in size, include prokaryotic cyanobacteria and eukaryotic algae. Nanoplankton, ranging from 2 to 20 micrometers, often consist of small flagellates and some types of dinoflagellates. Microplankton, greater than 20 micrometers, primarily include diatoms and larger dinoflagellates (Chisholm, 1992).

1.2 Factors Controlling Phytoplankton Size Structure in Coastal Water

1.2.1 Nutrient Availability

Nutrient availability is a key factor influencing phytoplankton size structure. In nutrient-rich environments, larger phytoplankton, such as diatoms, tend to dominate because they can efficiently utilize the abundant nutrients and grow rapidly. Conversely, in nutrient-poor conditions, smaller phytoplankton, such as picoplankton and small nanoplankton, are more competitive due to their higher surface area-to-volume ratio, which allows for more efficient nutrient uptake (Finkel et al., 2010).

1.2.2 Light Intensity

Light availability affects phytoplankton growth and size distribution. Smaller phytoplankton, such as picoplankton, are better adapted to low light conditions due to their efficient light absorption capabilities. In contrast, larger phytoplankton require higher light intensities to support their larger biomass. In coastal waters, where light penetration can be limited by turbidity and other factors, the size structure of phytoplankton communities can shift in response to changing light conditions (Irwin et al., 2015).

1.2.3 Temperature

Temperature plays a significant role in determining the metabolic rates and growth of phytoplankton. Different size classes of phytoplankton have varying thermal optima, with smaller phytoplankton generally thriving in warmer waters due to their higher metabolic rates. In Indian coastal waters, seasonal temperature variations can lead to shifts in phytoplankton size structure, with smaller phytoplankton becoming more prevalent during warmer periods (Marañón, 2015).

1.2.4 Grazing Pressure

Grazing by zooplankton and other herbivores is a major control on phytoplankton size structure (Kiørboe, 2011). Predation pressure can selectively remove larger phytoplankton from the community, allowing smaller phytoplankton to dominate. Additionally, the size and feeding preferences of grazers can influence the size distribution of phytoplankton. For example, copepods tend to feed on larger phytoplankton, while microzooplankton target smaller size classes (Calbet & Landry, 2004).

1.3 Case Studies and Observations in Indian Coastal Waters

Several studies have examined the size structure of phytoplankton in Indian coastal waters, highlighting the variability and complexity of these communities. For instance, Shunmugapandi et al. (2020) investigated the phytoplankton size distribution along the southwest coast of India and found that larger phytoplankton, particularly diatoms, dominated during the monsoon season due to increased nutrient inputs from river discharge and upwelling. In contrast, smaller phytoplankton, such as picoplankton and small flagellates, were more abundant during the pre-monsoon and post-monsoon periods when nutrient levels were lower.

Another study by Patil and Anil (2015) focused on the size structure of phytoplankton in the coastal waters off Goa. The researchers observed a seasonal shift in phytoplankton size classes, with larger phytoplankton, such as diatoms and dinoflagellates, dominating during the monsoon season and smaller phytoplankton, including picoplankton and nanoplankton, becoming more prevalent during the non-monsoon periods. These findings underscore the influence of nutrient availability and seasonal dynamics on the size structure of phytoplankton communities.

The impact of anthropogenic activities on phytoplankton size structure has also been studied in Indian coastal waters. For example, pollution from industrial and agricultural runoff can alter nutrient levels and availability, leading to shifts in phytoplankton size distribution. A study by Sahu et al. (2012) in the coastal waters off Chennai reported an increase in smaller phytoplankton, such as picoplankton and small nanoplankton, in areas affected by nutrient pollution. This shift was attributed to the higher surface area-to-volume ratio of smaller phytoplankton, which allows for more efficient nutrient uptake in polluted environments.

2 Major Functional Types of Phytoplankton

2.1 *Diatoms*

Diatoms are among the most ecologically significant groups of phytoplankton, particularly in nutrient-rich coastal waters. They possess siliceous cell walls, known as frustules, which are intricately patterned and contribute to their high sinking rates. Diatoms play a crucial role in primary production and are a key food source for zooplankton and other marine organisms. They are particularly dominant during upwelling events and periods of high nutrient availability (Sarthou et al., 2005). Recent studies have highlighted the importance of diatoms in the Indian coastal waters. For instance, D'Costa and Anil (2010) observed a seasonal abundance of diatoms in the coastal waters of Goa, correlating their presence with increased nutrient concentrations during the monsoon season. Similarly, a study by Gomes et al. (2000) reported high diatom diversity and biomass in the coastal waters of the Bay of Bengal, driven by riverine inputs and coastal upwelling.

2.2 *Dinoflagellates*

Dinoflagellates are another major group of phytoplankton characterized by their two flagella, which provide mobility. They can be autotrophic, heterotrophic, or mixotrophic, allowing them to adapt to a wide range of environmental conditions. Dinoflagellates are known for their ability to form harmful algal blooms (HABs), which can produce toxins that affect marine life and human health (Anderson et al., 2012). In Indian coastal waters, dinoflagellate blooms have been reported with increasing frequency. Sathish et al. (2022) documented the occurrence of a harmful dinoflagellate bloom along the southwest coast of India, highlighting the need for monitoring and management strategies to mitigate the impacts of such events. The study emphasized the role of anthropogenic activities and climate change in the proliferation of HABs.

2.3 *Coccolithophores*

Coccolithophores are small, unicellular algae that produce calcium carbonate plates, known as coccoliths. They play a significant role in the global carbon cycle and ocean acidification processes. The calcification process of coccolithophores contributes to the sequestration of carbon in the deep ocean (Rost & Riebesell, 2004). The northern Indian Ocean, highly vulnerable to climate change and acidification, lacks focused studies on coccolithophores, leaving their roles in productivity, calcite rain, carbon chemistry, and particulate inorganic carbon largely unknown (Arundhathy et al., 2021).

2.4 *Cyanobacteria*

Cyanobacteria, also known as blue-green algae, are prokaryotic phytoplankton capable of nitrogen fixation. This ability allows them to thrive in nutrient-poor environments where nitrogen is a limiting factor. Cyanobacteria are important contributors to primary production, particularly in oligotrophic (nutrient-poor) waters (Paerl, 2008). In Indian coastal waters, cyanobacteria play a vital role in sustaining primary production in areas with low nitrogen availability. Kumar et al. (2017) studied the distribution and ecological significance of cyanobacteria in the Arabian Sea, noting their contribution to nitrogen fixation and primary productivity. The study also discussed the impact of environmental changes on cyanobacterial dynamics and their potential role in biogeochemical cycles.

3 Environmental Factors Influencing Functional Types

3.1 *Temperature*

Temperature is a critical factor influencing the distribution and abundance of phytoplankton functional types. Different groups have varying thermal optima, affecting their metabolic rates and growth patterns. For instance, warmer temperatures tend to favor the growth of smaller phytoplankton, such as cyanobacteria and some dinoflagellates, while cooler temperatures are more conducive to the proliferation of diatoms (Marañón, 2015).

Recent observations in Indian coastal waters indicate that rising sea temperatures due to climate change are impacting phytoplankton communities. Rajalakshmi and Achyuthan (2021) noted that climate change–driven shifts in temperature, salinity, and ocean chemistry can modify phytoplankton community composition, thereby reshaping the marine food web and influencing productivity across higher trophic levels in the Nay of Bengal. These changes have implications for the entire marine food web and biogeochemical processes.

3.2 Salinity

Salinity variations, influenced by freshwater input and evaporation, also affect phytoplankton distribution. Coastal areas with significant freshwater influx, such as estuaries, often exhibit a diverse range of phytoplankton functional types adapted to varying salinity levels. Diatoms, for example, are generally more tolerant of lower salinities, while coccolithophores prefer more stable saline conditions (Smayda, 2002).

In Indian coastal waters, the influence of monsoonal rainfall and river discharge on salinity is pronounced. A study by Manna et al. (2010) in the Sundarbans estuarine system revealed the impact of seasonal salinity changes on phytoplankton community structure, with shifts from diatom dominance during low salinity periods to dinoflagellate and cyanobacteria dominance during high salinity periods.

3.3 Nutrient Availability

Nutrient availability, particularly nitrogen and phosphorus, is a primary driver of phytoplankton growth and functional diversity. Eutrophic conditions, often resulting from agricultural runoff and urban wastewater, can lead to blooms of opportunistic species such as dinoflagellates and cyanobacteria, while oligotrophic conditions favor smaller phytoplankton like picoplankton (Finkel et al., 2010).

Recent studies have highlighted the impact of nutrient loading on phytoplankton dynamics in Indian coastal waters. Tarafdar et al. (2023) examined the environmental drivers of Microcystis bloom (toxin-producing cyanobacteria) in Chilika Lagoon, identifying warm, low-salinity, P-limited conditions with high transparency as the key factors promoting bloom development. The study emphasized the need for nutrient management to prevent harmful algal blooms and maintain ecological balance.

3.4 Grazing Pressure

Grazing by zooplankton and other herbivores is a major factor influencing the size structure and functional composition of phytoplankton communities (Zhao et al., 2008). Grazing pressure can shape phytoplankton dynamics through selective predation, which can alter species composition, biomass, and the overall productivity of marine ecosystems.

Zooplankton, such as copepods, microzooplankton, and larger herbivores, exert selective pressure on phytoplankton communities by preferentially feeding on certain size classes and types of phytoplankton. This selective grazing can lead to shifts in the size structure and species composition of phytoplankton communities. For example, copepods typically prefer larger phytoplankton, such as diatoms, while microzooplankton tend to feed on smaller phytoplankton, including picoplankton and small flagellates (Calbet & Landry, 2004).

In Indian coastal waters, grazing pressure varies seasonally and spatially, influenced by factors such as temperature, nutrient availability, and predator-prey interactions (Ezhilarasan et al., 2025). Studies have shown that during periods of high grazing pressure, smaller phytoplankton may dominate due to the removal of larger cells by zooplankton. Conversely, during periods of low grazing pressure, larger phytoplankton can proliferate, leading to changes in community structure and productivity (Lancelot et al., 2009).

4 Anthropogenic Influences on Phytoplankton Communities

Human activities have increasingly influenced phytoplankton communities in coastal waters worldwide, including those in India (Halpern et al., 2008). This section explores the various anthropogenic factors affecting phytoplankton communities, including pollution, eutrophication, climate change, and coastal development, highlighting recent research findings and their implications.

4.1 Pollution

4.1.1 Nutrient Pollution and Eutrophication

Nutrient pollution, primarily from agricultural runoff and wastewater discharge, leads to eutrophication in coastal waters. Excessive nutrients, particularly nitrogen and phosphorus, promote the rapid growth of phytoplankton, often resulting in harmful algal blooms (HABs). These blooms can deplete oxygen levels, produce toxins, and disrupt marine ecosystems (Paerl, 2018).

Recent studies in Indian coastal waters have documented the impact of nutrient pollution on phytoplankton communities. Sahu et al. (2020) investigated the effects of nutrient enrichment in Chilika Lagoon, revealing an increase in cyanobacterial blooms linked to agricultural runoff. This shift in phytoplankton composition can reduce biodiversity and affect higher trophic levels, including fish populations.

4.1.2 Chemical Contaminants

Industrial discharge and urban runoff introduce various chemical contaminants into coastal waters, including heavy metals, pesticides, and pharmaceuticals. These pollutants can have direct toxic effects on phytoplankton or alter their community structure by affecting nutrient dynamics (Schernewski et al., 2025).

For instance, a study by Shahi et al. (2015) examined the presence of heavy metals in the coastal waters of Mumbai and their impact on phytoplankton. The researchers found that high concentrations of lead and mercury negatively affected diatom populations, leading to a dominance of more tolerant species such as certain dinoflagellates and cyanobacteria.

4.2 Climate Change

4.2.1 Temperature Changes

Global warming leads to increased sea surface temperatures, which can significantly affect phytoplankton communities. Temperature influences metabolic rates, growth, and the distribution of phytoplankton functional types. Warmer temperatures generally favor smaller phytoplankton, such as picoplankton and some cyanobacteria, over larger diatoms (Marañón, 2015).

Recent research by Sharma et al. (2021) in the Bay of Bengal documented shifts in phytoplankton composition due to rising sea temperatures. The study observed a decline in diatom abundance and an increase in cyanobacteria and small flagellates, which could alter the marine food web and biogeochemical cycles.

4.2.2 Ocean Acidification

Increased atmospheric CO_2 levels lead to ocean acidification, which affects calcifying organisms like coccolithophores. Acidification can reduce the availability of carbonate ions needed for coccolith formation, potentially decreasing their populations and impacting the carbon cycle (Rost et al., 2008).

Shetye et al. (2023) studied the effects of ocean acidification on coccolithophores in the northern Indian Ocean. The researchers found that lower pH levels resulted in reduced calcification rates and shifts in species composition, with potential implications for carbon sequestration and marine food webs.

4.3 Coastal Development

4.3.1 Habitat Modification

Coastal development, including land reclamation, construction of ports, and urbanization, leads to habitat modification and loss of natural habitats such as mangroves and seagrasses (Duarte et al., 2008). These changes can directly impact phytoplankton by altering nutrient dynamics, light availability, and water circulation patterns. A study by Gogoi et al. (2021) examined the impact of coastal development on phytoplankton communities in the Sundarbans estuarine system. The researchers observed that habitat modification led to increased turbidity and nutrient levels, promoting the growth of opportunistic phytoplankton species and reducing overall biodiversity.

4.3.2 Pollution from Coastal Industries

Industrial activities along coastlines contribute to the introduction of pollutants, including oil spills, heavy metals, and organic waste, which can have detrimental effects on phytoplankton communities. These pollutants can cause physiological stress or mortality in sensitive phytoplankton species, leading to shifts in community structure (Lancelot et al., 2007). For example, an investigation by Banerjee et al. (2020) in the coastal waters off Chennai found that industrial effluents significantly altered phytoplankton diversity and abundance. The study highlighted the prevalence of more tolerant species, such as certain cyanobacteria and dinoflagellates, in polluted areas.

4.4 Fisheries and Overexploitation

4.4.1 Nutrient Imbalances from Fisheries

Overfishing and the removal of top predators can cause trophic cascades, altering nutrient cycling and availability in coastal ecosystems. These changes can impact phytoplankton communities by modifying the competitive dynamics and availability of nutrients required for their growth (Griffiths et al., 2017).

A study by Bode et al. (2019) in the coastal waters of Kerala explored the effects of overfishing on nutrient dynamics and phytoplankton composition. The researchers found that reduced fish populations led to increased nutrient availability, favoring the growth of fast-growing phytoplankton species such as diatoms and cyanobacteria.

4.4.2 Aquaculture Practices

The expansion of aquaculture, particularly in coastal regions, can introduce excess nutrients and organic matter into the water, promoting eutrophication and algal blooms. These changes can significantly impact the composition and functioning of phytoplankton communities (Holmer et al., 2008).

In a study conducted by Penmetsa et al. (2013) in the coastal waters of Andhra Pradesh, the impact of aquaculture on phytoplankton was assessed. The findings indicated that nutrient inputs from aquaculture operations led to increased phytoplankton biomass and a shift towards more opportunistic species, such as certain cyanobacteria and dinoflagellates.

5 Strategies for Managing and Conserving Phytoplankton Communities

Effective management and conservation of phytoplankton communities are vital for maintaining marine ecosystem health (Levin et al., 2009). Several strategies can be implemented to achieve this goal:

5.1 Nutrient Management

Reducing nutrient inputs from agricultural runoff, sewage discharge, and industrial effluents is critical for preventing eutrophication and HABs. Implementing best management practices (BMPs) in agriculture, such as buffer strips and controlled fertilizer application, can significantly reduce nutrient runoff. Upgrading wastewater treatment facilities to remove nutrients before discharge into coastal waters is also essential (Smith et al., 2006).

5.2 Pollution Control

Monitoring and controlling pollutants, including heavy metals, pesticides, and pharmaceuticals, is necessary to protect phytoplankton communities. Regular monitoring of coastal waters for contaminants and enforcing regulations to limit industrial discharges can mitigate the adverse effects of pollution on phytoplankton diversity and health (Shekhar et al., 2008).

5.3 Marine Protected Areas (MPAs)

Establishing and effectively managing MPAs can help conserve critical habitats and protect phytoplankton diversity. MPAs can reduce human impacts (Lubchenco, 2013), such as overfishing and habitat destruction, allowing phytoplankton communities to thrive and maintain their ecological roles. Implementing MPAs requires stakeholder engagement, adequate enforcement, and regular monitoring to ensure their effectiveness (Edgar et al., 2014).

5.4 Climate Change Mitigation

Addressing climate change through mitigation strategies, such as reducing greenhouse gas emissions, is crucial for protecting phytoplankton communities from the adverse effects of warming and ocean acidification. International agreements, such as the Paris Agreement, aim to limit global temperature rise and promote actions to enhance climate resilience in marine ecosystems (IPCC, 2019).

5.5 Restoration Projects

Restoration of degraded coastal habitats, such as mangroves, seagrasses, and coral reefs, can enhance the resilience of phytoplankton communities. These habitats provide essential services, including nutrient cycling, carbon sequestration, and habitat for various marine organisms. Restoration efforts should focus on rehabilitating these ecosystems to support healthy and diverse phytoplankton populations (Duarte & Regaudie-de-Gioux, 2009).

6 Conclusion

In conclusion, the study of phytoplankton size structure and functional types in Indian coastal waters provides valuable insights into the ecological dynamics of these vital marine ecosystems. The ongoing environmental changes and human impacts in coastal ecosystems necessitate a thorough understanding of phytoplankton dynamics in terms of their size and functionality for informed management and conservation strategies, ensuring the health and sustainability of marine ecosystems for future generations.

Acknowledgements Mr. Manoj Kumar Das gratefully acknowledges the fellowship support from the SAC, ISRO, under the 'Oceansat-3 Utilization Program (EOS-06 OCM)' for the project titled 'Satellite Detection and Evaluation of Phytoplankton Size Classes During Recurring Bloom Events in Coastal Waters of the Northwestern Bay of Bengal'.

References

Anderson, D. M., Cembella, A. D., & Hallegraeff, G. M. (2012). Progress in understanding harmful algal blooms: Paradigm shifts and new technologies for research, monitoring, and management. *Annual Review of Marine Science, 4*, 143–176. https://doi.org/10.1146/annurev-marine-120308-081121

Arundhathy, M., Jyothibabu, R., Santhikrishnan, S., Albin, K. J., Parthasarathi, S., & Rashid, C. P. (2021). Coccolithophores: An environmentally significant and understudied phytoplankton group in the Indian Ocean. *Environmental Monitoring and Assessment, 193*(3), 144.

Banerjee, P., Das, A., & Choudhury, A. (2020). Impact of industrial effluents on phytoplankton diversity and abundance in the coastal waters off Chennai, India. *Marine Pollution Bulletin, 156*, 111231. https://doi.org/10.1016/j.marpolbul.2020.111231

Behrenfeld, M. J., Boss, E., Siegel, D. A., & Shea, D. M. (2005). Carbon-based ocean productivity and phytoplankton physiology from space. *Global Biogeochemical Cycles, 19*(1). https://doi.org/10.1029/2004GB002299

Bode, M., Bode, L., & Armsworth, P. R. (2019). Overfishing alters trophic cascades and nutrient dynamics in marine ecosystems. *Nature Communications, 10*, 3126. https://doi.org/10.1038/s41467-019-11025-z

Boyd, P. W., Strzepek, R., Fu, F., & Hutchins, D. A. (2010). Environmental control of open-ocean phytoplankton groups: Now and in the future. *Limnology and Oceanography, 55*(3), 1353–1376.

Calbet, A., & Landry, M. R. (2004). Phytoplankton growth, microzooplankton grazing, and carbon cycling in marine systems. *Limnology and Oceanography, 49*(1), 51–57.

Chisholm, S. W. (1992). Phytoplankton size. In *Primary productivity and biogeochemical cycles in the sea* (pp. 213–237).

D'Costa, P. M., & Anil, A. C. (2010). Diatom community dynamics in a tropical, monsoon-influenced environment: West coast of India. *Continental Shelf Research, 30*(12), 1324–1337.

Duarte, C. M., & Regaudie-de-Gioux, A. (2009). Thresholds of gross primary production for the metabolic balance of marine planktonic communities. *Limnology and Oceanography, 54*(3), 1015–1022.

Duarte, C. M., Borum, J., Short, F. T., & Walker, D. I. (2008). Seagrass ecosystems: Their global status and prospects. In *Aquatic ecosystems: Trends and global prospects* (pp. 281–294). Cambridge University Press.

Edgar, G. J., Stuart-Smith, R. D., Willis, T. J., Kininmonth, S., Baker, S. C., Banks, S., et al. (2014). Global conservation outcomes depend on marine protected areas with five key features. *Nature, 506*(7487), 216–220. https://doi.org/10.1038/nature13022

Ezhilarasan, P., Panda, U. S., Pandiya Rajan, R. S., Vijay, A., Kumaraswami, M., Jinoj, T. P. S., & Murthy, M. R. (2025). Zooplankton community changes across the Coringa mangrove and adjacent waters of east coast of India: Insights into habitat-specific interactions with environmental changes. *Environmental Monitoring and Assessment, 197*(7), 826.

Falkowski, P. (2012). Ocean science: The power of plankton. *Nature, 483*(7387), S17–S20.

Finkel, Z. V., Beardall, J., Flynn, K. J., Quigg, A., Rees, T. A. V., & Raven, J. A. (2010). Phytoplankton in a changing world: Cell size and elemental stoichiometry. *Journal of Plankton Research, 32*(1), 119–137. https://doi.org/10.1093/plankt/fbp098

Friedland, K. D., Stock, C., Drinkwater, K. F., Link, J. S., Leaf, R. T., Shank, B. V., et al. (2012). Pathways between primary production and fisheries yields of large marine ecosystems. *PLoS One, 7*(1), e28945. https://doi.org/10.1371/journal.pone.0028945

Giering, S. L. C., Sanders, R., Lampitt, R. S., Anderson, T. R., Tamburini, C., & Boutrif, M. (2014). Reconciliation of the carbon budget in the ocean's twilight zone. *Nature, 507*(7493), 480–483. https://doi.org/10.1038/nature13123

Gogoi, P., Das, S. K., Sarkar, S. D., Chanu, T. N., Manna, R. K., Sengupta, A., et al. (2021). Environmental factors driving phytoplankton assemblage pattern and diversity: Insights from Sundarban eco-region, India. *Ecohydrology & Hydrobiology, 21*(2), 354–367.

Gomes, H. R., Goes, J. I., & Saino, T. (2000). Influence of physical processes and freshwater discharge on the seasonality of phytoplankton regime in the Bay of Bengal. *Continental Shelf Research, 20*(3), 313–330.

Griffiths, J. R., Kadin, M., Nascimento, F. J., Tamelander, T., Törnroos, A., Bonaglia, S., et al. (2017). The importance of benthic–pelagic coupling for marine ecosystem functioning in a changing world. *Global Change Biology, 23*(6), 2179–2196. https://doi.org/10.1111/gcb.13642

Hallegraeff, G. M. (1993). A review of harmful algal blooms and their apparent global increase. *Phycologia, 32*(2), 79–99. https://doi.org/10.2216/i0031-8884-32-2-79.1

Halpern, B. S., Walbridge, S., Selkoe, K. A., Kappel, C. V., Micheli, F., D'Agrosa, C., et al. (2008). A global map of human impact on marine ecosystems. *Science, 319*(5865), 948–952. https://doi.org/10.1126/science.1149345

Holmer, M., Argyrou, M., Dalsgaard, T., Danovaro, R., Diaz-Almela, E., Duarte, C. M., et al. (2008). Effects of fish farm waste on Posidonia oceanica meadows: synthesis and provision of monitoring and management tools. *Marine Pollution Bulletin, 56*(9), 1618–1629.

IPCC. (2019). *The ocean and cryosphere in a changing climate: Summary for policymakers*. Intergovernmental Panel on Climate Change. https://www.ipcc.ch/srocc/

Irwin, A. J., Finkel, Z. V., Müller-Karger, F. E., & Troccoli Ghinaglia, L. (2015). Phytoplankton adapt to changing ocean environments. *Proceedings of the National Academy of Sciences, 112*(18), 5762–5766. https://doi.org/10.1073/pnas.1414752112

Kiørboe, T. (2011). How zooplankton feed: Mechanisms, traits and trade-offs. *Biological Reviews, 86*(2), 311–339. https://doi.org/10.1111/j.1469-185X.2010.00148.x

Kumar, P. K., Singh, A., Ramesh, R., & Nallathambi, T. (2017). N_2 fixation in the eastern Arabian Sea: Probable role of heterotrophic diazotrophs. *Frontiers in Marine Science, 4*, 80.

Lancelot, C., Gypens, N., Billen, G., Garnier, J., & Roubeix, V. (2007). Testing an integrated river–ocean mathematical tool for linking marine eutrophication to land use: The Phaeocystis-dominated Belgian coastal zone (Southern North Sea) over the past 50 years. *Journal of Marine Systems, 64*(1–4), 216–228.

Lancelot, C., Rousseau, V., & Gypens, N. (2009). Ecologically based indicators for Phaeocystis disturbance in eutrophied Belgian coastal waters (Southern North Sea) based on field observations and ecological modelling. *Journal of Sea Research, 61*(1–2), 44–49.

Levin, P. S., Fogarty, M. J., Murawski, S. A., & Fluharty, D. (2009). Integrated ecosystem assessments: Developing the scientific basis for ecosystem-based management of the ocean. *PLoS Biology, 7*(1), e14. https://doi.org/10.1371/journal.pbio.1000014

Lubchenco, J. (2013). Large marine ecosystems: The leading edge of science, management and policy. In *Stress, sustainability, and development of large marine ecosystems during climate change: Policy and implementation* (pp. 2–19).

Manna, S., Chaudhuri, K., Bhattacharyya, S., & Bhattacharyya, M. (2010). Dynamics of Sundarban estuarine ecosystem: Eutrophication induced threat to mangroves. *Saline Systems, 6*(1), 8.

Marañón, E. (2015). Cell size as a key determinant of phytoplankton metabolism and community structure. *Annual Review of Marine Science, 7*(1), 241–264. https://doi.org/10.1146/annurev-marine-010814-015955

Paerl, H. (2008). Nutrient and other environmental controls of harmful cyanobacterial blooms along the freshwater–marine continuum. In *Cyanobacterial harmful algal blooms: State of the science and research needs* (pp. 217–237). Springer.

Paerl, H. W. (2018). Mitigating toxic planktonic cyanobacterial blooms in aquatic ecosystems facing increasing anthropogenic and climatic pressures. *Toxins, 10*(2), 76.

Patil, J. S., & Anil, A. C. (2015). Effect of monsoonal perturbations on the occurrence of phytoplankton blooms in a tropical bay. *Marine Ecology Progress Series, 530*, 77–92.

Penmetsa, A. R. K. R., Muppidi, S. R., Popuri, R., Golla, S. B., & Tenneti, R. (2013). Impact of aquaculture on physico-chemical characteristics of water and soils in the coastal tracts of East and West Godavari Districts, Andhra Pradesh, India. *International Journal of Engineering Trends and Technology (IJETT), 6*(6), 313–319.

Rajalakshmi, P. R., & Achyuthan, H. (2021). Climate change as observed in the Bay of Bengal. *Journal of Climate Change, 7*(3), 69–82.

Rost, B., & Riebesell, U. (2004). Coccolithophores and the biological pump: Responses to environmental changes. In *Coccolithophores: From molecular processes to global impact* (pp. 99–125). Springer.

Rost, B., Zondervan, I., & Wolf-Gladrow, D. (2008). Sensitivity of phytoplankton to future changes in ocean carbonate chemistry: Current knowledge, contradictions and research directions. *Marine Ecology Progress Series, 373*, 227–237.

Sahu, G., Satpathy, K. K., Mohanty, A. K., & Sarkar, S. K. (2012). Variations in community structure of phytoplankton in relation to physicochemical properties of coastal waters, southeast coast of India. *Indian Journal of Geo-marine Sciences, 41*(3), 223.

Sahu, G., Pati, P., & Behera, P. (2020). Impact of nutrient enrichment on phytoplankton communities in Chilika lagoon, India. *Marine Pollution Bulletin, 157*, 111291. https://doi.org/10.1016/j.marpolbul.2020.111291

Sarthou, G., Timmermans, K. R., Blain, S., & Tréguer, P. (2005). Growth physiology and fate of diatoms in the ocean: A review. *Journal of Sea Research, 53*(1–2), 25–42. https://doi.org/10.1016/j.seares.2004.01.007

Sathish, T., Nazrin, A. K., Thomas, L. C., & Padmakumar, K. B. (2022). Seasonal dynamics of dinoflagellates with special emphasis on potentially harmful species in a tropical estuarine system along the southwest coast of India. *Journal of Oceanography, 78*(5), 397–408.

Schernewski, G., Schneider, M., Neumann, T., & von Weber, M. (2025). Phytoplankton dynamics in a large lagoon: Nutrient load reductions, climate change, and cold-and heatwaves. *Environments, 12*(10), 370.

Shahi, N., Godhe, A., Mallik, S. K., Härnström, K., & Nayak, B. B. (2015). The relationship between variation of phytoplankton species composition and physico-chemical parameters in northern coastal waters of Mumbai, India. *Indian Journal of Geo-Marine Sciences, 44*(5), 673–684.

Sharma, R., Rao, P., & Pandey, S. (2021). Shifts in phytoplankton composition in response to rising sea temperatures in the Bay of Bengal. *Marine Biology, 168*(6), 75. https://doi.org/10.1007/s00227-021-03893-1

Shekhar, T. S., Kiran, B. R., Puttaiah, E. T., Shivaraj, Y., & Mahadevan, K. M. (2008). Phytoplankton as index of water quality with reference to industrial pollution. *Journal of Environmental Biology, 29*(2), 233.

Shetye, S., Gazi, S., Manglavil, A., Shenoy, D., Kurian, S., Pratihary, A., et al. (2023). Malformation in coccolithophores in low pH waters: Evidences from the eastern Arabian Sea. *Environmental Science and Pollution Research, 30*(14), 42351–42366.

Shunmugapandi, R., Inamdar, A. B., & Gedam, S. K. (2020). Long-time-scale investigation of phytoplankton communities based on their size in the Arabian Sea. *International Journal of Remote Sensing, 41*(15), 5992–6009.

Smayda, T. J. (2002). Adaptations and selection of harmful and other dinoflagellate species in upwelling systems. *Journal of Marine Systems, 3*(4), 613–639. https://doi.org/10.1016/S0924-7963(01)00067-7

Smith, V. H., Joye, S. B., & Howarth, R. W. (2006). Eutrophication of freshwater and marine ecosystems. *Limnology and Oceanography, 51*(1part2), 351–355.

Tarafdar, L., Mohapatra, M., Muduli, P. R., Kumar, A., Mishra, D. R., & Rastogi, G. (2023). Co-occurrence patterns and environmental factors associated with rapid onset of Microcystis aeruginosa bloom in a tropical coastal lagoon. *Journal of Environmental Management, 325*, 116580.

Zhao, J., Ramin, M., Cheng, V., & Arhonditsis, G. B. (2008). Plankton community patterns across a trophic gradient: The role of zooplankton functional groups. *Ecological Modelling, 213*(3–4), 417–436. https://doi.org/10.1016/j.ecolmodel.2008.01.016

Chapter 7
Bio-Optical Properties of Hugli Estuary: A Review

Rituparna Acharyya, Michal Habel, Niloy Pramanick, Jyoti Prakash Hati, Kaberi Samanta, Sarmistha Basu, and Anirban Mukhopadhyay

Abstract The Hugli estuary, a distributary of the Ganges extending between 21°32′–22°40′ N latitude and 88°05′–89°00′ E longitude, is primarily influenced by tides and right-bank tributaries, with limited freshwater input from the Farakka barrage. The estuary is characterized as "well-mixed" due to its meso-macro tidal setting and semi-diurnal tidal regime, with significant tidal amplitudes. Flow dynamics are driven by the prevailing monsoonal climate, which brings substantial seasonal variation in river discharge. Satellite data, like Sentinel-2 MSI, help assess TSM Concentration (CTSM), which varies between 132 and 540 mg/I during the pre-monsoon season. CDOM levels are higher during monsoons and near river confluences, decreasing offshore. Seasonal variations show higher Chl-a during monsoons due to nutrient flux, with concentrations ranging from 2.96 to 3.17 mg m^{-3}, highlighting productive fishing zones along the continental shelf. The Hugli estuary's water temperature varies by 10 °C seasonally, influencing water density and light absorption. During post-monsoons, temperatures range between 27 and 28°, affecting mixing and circulation patterns. Salinity levels range from 4.5 to 13 ppt during the dry season and drop significantly post-monsoon. The Farakka barrage upstream has altered salinity patterns, affecting organism distribution. In the Hugli estuary, DO levels peak post-monsoon and are lower pre-monsoon, showing an inverse relationship with water temperature. Potential Fishing Zones (PFZs) are identified using bio-optical properties like salinity, SST, and Chl-a. Studies indicate a strong correlation between nutrient-rich waters, phytoplankton biomass, and fish aggregation, essential for sustainable fishery management in the Hugli estuary.

Keywords Hugli Estuary · Bio-optical · TSA · CDOM · CHL · PFZ

R. Acharyya · M. Habel · J. P. Hati · A. Mukhopadhyay (✉)
Faculty of Geographical Sciences, Kazimierz Wielki University, Bydgoszcz, Poland

N. Pramanick
Science for Sustainability, Kolkata, India

K. Samanta · S. Basu
Calcutta Institute of Engineering and Management, Kolkata, India

© The Author(s), under exclusive license to Springer Nature Switzerland AG 2026

S. Das (ed.), *Bio-optical Properties of Indian Coastal Waters*,
https://doi.org/10.1007/978-3-032-08714-0_7

1 Introduction

1.1 The Dynamic Hugli Estuary and Its Importance

The estuary of Bhagirathi-Hugli River system, i.e., Hugli estuary, extending between 21°32′ and 22°40′ North latitude; 88°05′ and 89°00′ East longitude, is a distributary of the River Ganges flowing into the Bay of Bengal and constitutes the major artery for the Sundarbans mangroves (De et al., 2015). However, this river system has deteriorated throughout history, as documented in various studies (Sherwill, 1858; Reaks, 1919; Morgan & McIntire, 1959; Bagchi, 1944; Basu & Chakraborty, 1972). Hence, in its lower course, the Hugli estuary is primarily maintained by tides and right-bank tributaries, except the Farakka barrage project, situated approximately 286 km upstream, which diverts a sizable discharge from the Ganges (Bandyopadhyay, 2000). It acts as an open estuary, year-round, receiving certain amount of freshwater discharge from the Ganges, which flows continuously via the Farakka barrage (Akhand et al., 2016; Biswas et al., 2007). However, the maximum discharge from the barrage, up to 1135 $m^3\ s^{-1}$, can only raise the ebb fluxes at the estuary's sea face by about 2%, with limited impact on the flow conditions of the estuary (Roy, 1969). As a result, the Hugli estuary's flow is now mainly controlled by tides and its right-bank tributaries (e.g., Rupnarayan, Haldi, and Rasulpur rivers).

The Hugli estuary is known as a "well-mixed" estuary because of its meso-macro tidal setting, with amplitudes between 2.5 and 7.0 m (De et al., 2015). The Hugli estuary possesses a predominant semi-diurnal tidal regime because of its funnel-shaped mouth (De et al., 2015). Hence, it is an estuarine complex, with a range of spring tides between 4.27 and 4.75 m and neap tides between 1.83 and 2.83 m. The flood current across most of this estuary ranges between 0.5 and 2.0 $m\ s^{-1}$ (Biswas et al., 2007), with average current velocities ranging between 117 and 108 $cm\ s^{-1}$, respectively, during ebb and flood tides (De et al., 2011). The flood tide lasts approximately 3 h during a tidal cycle of 12.4 h, resulting in flood velocities between 2 and 3 $m^3\ s^{-1}$ (Sanyal & Chakrabarti, 1995). At Hugli estuary's mouth, the maximal flood and ebb discharge amounts to 2.6 × 105 $m^3 s^{-1}$ and 1.09 × 105 $m^3 s^{-1}$, respectively (McDowell, 1995) (Fig. 7.1).

The prevailing monsoonal climate (southwest monsoon from July to October, northeast or post-monsoon from November to February, and pre-monsoon from March to June) significantly influences the flow dynamics of the Hugli estuary. The southwest monsoon brings 70–80% of the precipitation and causes high river discharge ranging between 2952 and 11,897 $m^3\ s^{-1}$, which decreases to range 900–1500 $m^3\ s^{-1}$ during the non-monsoonal months (Biswas et al., 2004; Mukhopadhyay et al., 2006). The Hugli Estuary also exhibits dynamics in case of depth variations ranging from around ~21 m in Diamond Harbor to around ~8 m near the estuary's mouth (Bhaumik & Sharma, 2012). The Hugli estuary can be subdivided morphologically into flood- and ebb-dominated pathways that are mutually evasive and bordered by linear tidal sandy ridges (Roy, 1969). The monsoonal climate also influences the Hugli estuary's sedimentation scenario (Sarkar et al.,

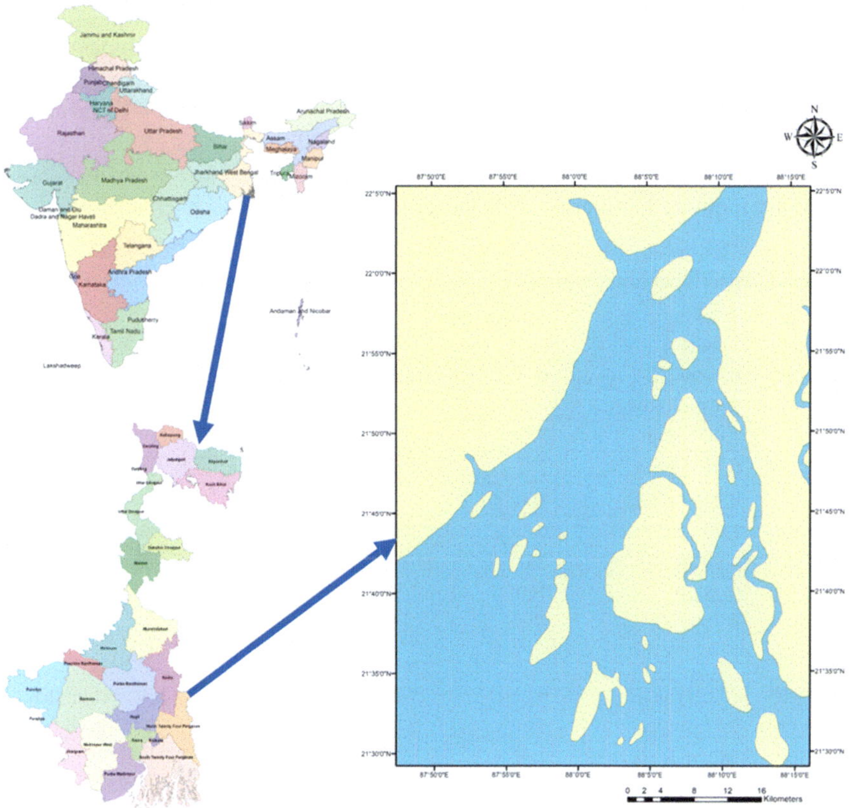

Fig. 7.1 The Hugli estuary at the confluence of the Hugli River at the Bay of Bengal

2005). Most sedimentation occurs during the pre-monsoon and post-monsoon, and its sizeable portion is flushed out during the monsoonal peak flow (Mukhopadhyay et al., 2023; Sarkar et al., 2005). The estuary exhibits varying sediment textures, with coarser sediments of 0.12 mm mean grain size (D50) in the upper section and finer sediments of 0.09 mm in the lower section (Arora & Bhaskaran, 2013).

1.2 Properties of Water and Estuarine Water

1.2.1 Properties of Water

Water's properties are significant in various scientific fields, including chemistry, physics, geology, and climate studies. Despite decades of investigation, water dynamics across diverse conditions and ecosystems remain perplexing and often intriguing (Paesani & Voth, 2009). Water is a simple molecule with only three atoms although its unique properties render it the most significant molecule in existence (Abu Shmeis, 2018). Water molecules constitute a 3-D network, comprising each atom of oxygen tetrahedrally connected to four atoms of hydrogen via two covalent and hydrogen bonds, which renders water its distinct physical (Table 7.1) and chemical features (Ebbing & Gammon, 1996; Manahan, 2022).

Understanding environmental processes that incorporate water as an element of reactions requires familiarity with the critical chemical characteristics of water

Table 7.1 Water's physical properties

Physical property	Values	Temperature (°C)	Effect and importance
Density	0.9999 g/mL (gram per milliliter	0	Water exists in three physical states: solid (ice), liquid, and gas (water vapor). Ice has less density than liquid water, causing it to float. Water vapor possesses the lowest density, compared to other states
	1.000 g/mL	3.95	
	0.9971 g/mL	25	
	0.9584 g/mL	100	
Viscosity	1.79 cp (Poise)	0	Water viscosity quantifies the resistance to flow caused by molecular interactions. Water viscosity decreases as temperature rises because water particles gain thermal energy, which reduces molecular interactions and resistance to flow
	0.89 cP	25	
	0.28 cP	100	
Surface tension	75.6 dyn/cm	0	Water's surface tension may hold weight that might otherwise sink. Surface tension allows certain insects to move across it effortlessly and plays a vital role in creating waves essential for transporting oxygen in bodies of water. Notably, besides mercury, water has the highest surface tension among commonly occurring liquids
	72.5 dyn/cm	25	
	58.9 dyn/cm	100	
Dipole moment	1.8546 D (Debeye)	–	Surface tension in water may prevent weight from sinking. For instance, surface tension also allows certain insects (striders) to move effortlessly across water. It plays a vital role in creating waves essential for transporting oxygen in bodies of water. The surface tension of water is the highest among commonly occurring liquids, aside from mercury
Specific heat	1 cal/g/°C	–	Water can absorb significant amounts of heat while very significantly increasing its temperature
Turbidity	Low: 10 NTUs (Nephelometric turbidity units): Treated water	–	Turbidity serves as an indirect indicator of the quantity of suspended particles in the water. It rises with the concentration of suspended solids within the water and decreases the depth to which light can penetrate
	Moderate: 50 NTUs		
	High: >100 NTUs		

Source: (Snoeyink & Jenkins, 1991; Ebbing & Gammon, 1996; Hanrahan, 2012; Abu Shmeis, 2018; DataStream Initiative, 2021; Manahan, 2022)

(H_2O) (Mudryk & Mitchenko, 2023). Water's amphoteric nature allows it to operate as both an acid and a base; it contains hydrogen, allowing it to act as an acid, and its oxygen atom can also form a bond with hydrogen, enabling it to act as a base (Abu Shmeis, 2018). Water is essential for redox reactions because it can be oxidized and reduced. The oxidation of a water molecule during photosynthesis is one of the most critical processes (Brudvig et al., 1989). Water plays a crucial role as a universal solvent in various processes. By its very nature, it seldom exists in its purest form, often containing gases, inorganic salts, and organic compounds (Mudryk & Mitchenko, 2023). Water hardness occurs from the accumulation of minerals when water passes through rocks, enabling specific ions to dissolve in water based on their solubility and composition. For example, minerals present in rocks, such as limestone ($CaCO_3$), gypsum ($CaSO_4$), and dolomite ($MgCO_3$) contribute to water hardness (Mudryk & Mitchenko, 2023). The amount of dissolved minerals in water can help identify its source. Groundwater is more complex than surface water due to prolonged contact with rocks and minerals. Water hardness can reduce metal toxicity to aquatic organisms. Water hardness is often quantified as equivalents of calcium carbonate (mg/L $CaCO_3$). However, there are several other hardness scales and measuring units (DataStream Initiative, 2021). Table 7.2 depicts the hardness values.

Dissolved Oxygen (DO), a critical chemical property of water, which is typically measured in mg/l or ppm using modern techniques like electrochemical gas or optical sensors. The DO level in water is a crucial indicator of its health. Running water, for instance, benefits from oxygen from the atmosphere and aquatic plants, leading to higher oxygen levels than standing water (Onojake et al., 2023). However, excessive organic matter in water bodies can decrease the DO level, leading to issues like algal bloom, reduced oxygen levels, and harm to aquatic animals (DataStream Initiative, 2021).

pH is a chemical property that determines whether water is acidic or basic (e.g., lemon juice vs. bleach or detergent). It is measured on a scale from 0 (highly acidic) to 14 (strongly basic). A pH level of 7 is considered neutral, similar to that of pure water. Water's pH range is rarely <4 or >10; hence, assessing its pH is critical because it influences many biological and chemical activities (Davis & Cornwell, 1991).

Table 7.2 Water hardness values

Hardness values (mg/L $CaCO_3$)			
Soft	Moderately hard	Hard	Very hard
0–60	61–120	121–180	>180

Source: DataStream Initiative (2021)

1.2.2 Properties of Estuarine Water

The above-mentioned physical and chemical properties of water discussed briefly. However, the present study centers on estuaries, which are dynamic systems wherein the freshwater-saltwater ratio varies depending on the tide, freshwater fluxes, seasonality, and the opening of the estuary's mouth. These properties influence the variation in estuarine dynamics, especially the estuaries with depth, called depth profile, which can exhibit significant variations between its upper reaches and its connection to the sea (Estuary Watch, 2011). Regular monitoring of the properties of water within estuaries is essential, as it leads to creating a dataset, which helps determine which processes may occur within the estuary at different periods of the year (Wenner, 2001). Regular monitoring helps assess how human activities induce fluctuations in natural processes operating across the estuaries. This approach also contributes to understanding the types of habitats within the estuary and how the changes in the estuarine environment impact their survival (Estuary Watch, 2011). The most significant properties of estuarine water are Dissolved Oxygen (DO) concentration, salinity, pH, and turbidity.

The concentration of Dissolved Oxygen (DO) within any estuary is determined by the balance of mechanical processes that transfer oxygen-rich water through estuaries, including advection of freshwater upstream, vertical mixing of water, gravitational circulation, and biochemical mechanisms, including respiration, photosynthesis, and organic decomposition that generate and use oxygen (Shen & Qin, 2024). DO is critical for estuarine habitat health and diversity since they could perish or relocate if DO concentrations are insufficient (Thrush et al., 2014). In a stratified estuary, the freshwater layer may occasionally overlap with the saltwater layer (NOAA, 2021). In this scenario, organisms and bacteria in the bottom layer of the estuary may consume all the oxygen, resulting in fish die-offs, especially if the surface layer is abruptly drained (Estuary Watch, 2011).

Salinity is critical for aquatic flora and fauna adapted to survive in specific salinity ranges within an estuary, which is determined by its shape, size, and the amount of fresh and saltwater entering and leaving it (McLusky, 1989). Estuarine salinity is mainly determined by how freshwater and marine fluxes interact; thus, the salinity range can differ significantly with the tide, from 35 ppt at high tide to 5 ppt at low tide with low freshwater intake (Estuary Watch, 2011). Generally, in estuaries, salinity peaks at its mouth and decreases upstream. In estuaries with minimal freshwater flow, evaporation can result in higher salinity (up to 60%) than seawater, forming stratified freshwater and saline water layers (NOAA, 2021). During certain times, waters within estuaries can blend, resulting in a homogenous water body. The timing and occurrence of these patterns impact the types of plants and animals that can thrive in the estuary (McLusky, 1989).

pH is a significant property for estuaries, as estuarine pH values typically range between 7.0 and 7.5 in freshwater section, and 8.0 and 8.6 in saline sections with higher salinity. Saline water in estuaries with these pH values (8.0–8.6) indicate slight alkalinity owing to the natural buffering of bicarbonate and carbonate dissolving (EPA, 2006). The pH levels within an estuary can fluctuate diurnally due to photosynthesis by aquatic plants, which can directly impact CO_2 concentrations (Raven et al., 2020). The pH level could also be reduced to approximately 5.5 by elevated rainfall through formation of sulfate acid from soil runoff (Virtasalo et al., 2023). Fluctuations in pH level could potentially impact both microbial activities and in turn, aquatic organisms. Extreme pH levels can harm fish organs, making them more susceptible to infections, while low pH levels can increase the toxicity of metals like aluminum (Gensemer & Playle, 1999).

The physicochemical properties of estuarine water dictate interaction of light in the water column and help to perceive water quality and ecosystem health. The following section provides a brief description of bio-optical properties of water.

2 Optics of Water and Saline Water

2.1 Optics of Water

In the visible spectrum of electromagnetic radiation, water appears transparent, yet attenuation and refraction of incoming light in the water column is a common phenomenon. The optical properties of water are dependent on interaction of water particles with incident light (Tilley, 2020). The refraction of light in water column is dependent on the incident angle, and the refractive index is determined by the temperature and salinity of water. The refractive index shows a proportional relationship with salinity and inverse relationship with temperature. The salinity of water at a constant temperature could be estimated from this relationship, if refractive index is already known (Lide, 2004; Papadopoulos et al., 2006; Fox, 2010; Alrifai et al., 2019).

The incident radiation on the ocean surface is either reflected or absorbed. The absorbed energy is converted to heat which can cause water to evaporate or warm up and can be used by plants for photosynthesis. Scattered light and reflected particles ascribe the ocean's color when viewed from above. The combination of water molecules, organic materials, dissolved salts, and suspended particles reduces the intensity of solar radiation at depths, demonstrating that solar radiation intensity drops with depth in ocean waters (Abdelly et al., 2022).

Both short (ultraviolet) and long (infrared) wavelengths are rapidly absorbed and cannot be diffused. Thus, only blue-green wavelengths can travel to any depth; the seas appear blue from above, so blue-green light scatters the most. Ocean waters can change color due to suspended or dissolved particles and variations in solar radiation quality caused by atmospheric conditions and the sun's angle. The purest ocean waters are located at a depth of 150 m, where only 1% of surface radiation is present. As a result, sunlight is not visible beyond 1000 m depth (Aghaleh et al., 2011; Annunziata et al., 2017; Alrifai et al., 2019) (Fig. 7.2).

Ocean light attenuation can be measured using various methods. A frequently used technique is a Secchi disk, a weighted circular white disk of around 30 cm

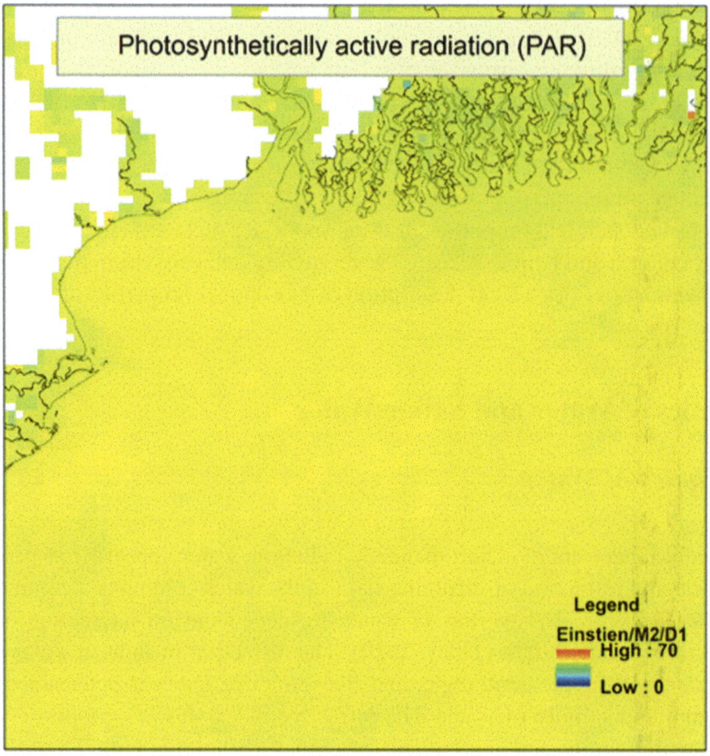

Fig. 7.2 PAR over the estuarine water of the Bay of Bengal

(12 in.) in diameter. It is immersed until the intensity of light backscattered from the water meets its reflectance (Suresh et al., 2006). The attenuation coefficient for available light averaged over the Secchi disk depth can be obtained by dividing this depth by 1.7 m. This allows the estimation of light intensity at a specific depth (I_z) from the intensity of light at the ocean surface (I_0) using Beer's formula in the form of $I_z = I_0 e^{xz}$, which considers the light extinction coefficient (Araus et al., 2021).

2.2 *Optics of Saline Water*

Salinity increases water's refractive index at a specific wavelength and temperature. However, the refractive index decreases at a certain salinity level as temperature rises. This indicates that the saline solution becomes optically less dense as the temperature increases, allowing light waves to travel more quickly. The intrinsic optical characteristics of ocean waters include total absorption and backscattering coefficients, which can further be divided into components such as particle backscattering, absorption due to phytoplankton, and absorption due to dissolved organic matter and detritus (Asseng et al., 2020; Atkinson & Bingman, 1997; Ben Hassine et al., 2009).

2.3 Physics of Light–Water Interaction

The most significant effect occurs when light strikes the water surface at a 45-degree angle. This effect is most pronounced with a specific type of polarization known as transverse magnetic polarization, and it reaches its peak in green light. Interestingly, green light is the color for which water is most transparent, resulting in the least interaction.

2.4 Light and Saline Water Interaction

Sun's electrical field affects molecules in the medium through which electromagnetic waves travel. Sunlight is the primary energy source for natural activities in the marine environment, providing electromagnetic radiation. Solar radiation entering the water is transformed into heat and other energy forms. The term "absorption spectrum" describes the relationship between light wavelength and the amount of energy a substance absorbs. The differences in absorption spectra in various seas indicate differences in the properties of the water in those seas. Absorption spectra can indirectly reveal the biological, geochemical, and thermodynamic processes that cause variations in oceanic waters. Other than absorption, scattering is also an important phenomenon in saline water which depends on interaction of incident light with suspended particles in water column (Breckle, 2002).

2.5 Estimation of Water Properties from Optics

Estimation of optical properties in estuarine water is crucial as they exhibit essential information on physical, chemical, and biological properties (Giardino et al., 2019; Mobley, 1995). The fate of the incident radiation on the water column (reflectance, absorption, or scattering) is primarily regulated by suspended sediments, dissolved or particulate organic and inorganic matter, and phytoplankton. The absorption coefficient is a measure of absorbed incident light by particles present in water whereas the scattering coefficient represent the amount of light dispersed by water particles (Bukata et al., 2018). Some amount of incident light is returned to the source through dispersion which is termed as backscattering. The attenuation coefficient includes the total loss of light through absorption and scattering phenomena (Mitchell et al., 2000; Haltrin, 2006). In the modern era, satellite-based measurements are more efficient than traditional mechanical sampling for measurement and monitoring of coastal waters as remotely sensed data provides synoptic and repetitive coverage. With the advanced optical sensors, it is possible to estimate underwater features such as sediment concentration, water quality, phytoplankton concentration, dissolved and particulate matter (Bricaud & Stramski, 1990; Sathyendranath, 2000; Miller et al., 2005).

3 Importance of Studying Bio-Optical Properties of Water

Bio-optical properties of water are crucial for comprehending and preserving aquatic ecosystems like lakes, rivers, estuaries, and coastal areas. These bio-optical properties dictate the interaction between marine organisms and surrounding water. Moreover, these properties could be used as an indicator for climate studies, fisheries management, detecting algal blooms, and monitoring water quality. Recent advances in sensors and algorithms, together with modeling are often used for informed decision-making by the stakeholders and governmental bodies for efficient management and preservation of water resources (Nayak, 2004; Kratzer et al., 2017).

3.1 Bio-Optical Properties of Hugli Estuary

3.1.1 Sediments

The bio-optical properties of sediments in the Hugli estuary include interaction of incident light with floating and settling materials in the dynamic estuarine water. Among the optically active compounds, total suspended matter (TSM) is the single most important parameter as it influences transport of nutrient and minerals, transmission of light, sedimentation, and pollution in water bodies (Venkatachalapathy et al., 2021). Concentration of TSM (C_{TSM}) is often influenced by anthropogenic properties and since it is heterogeneous and temporally dynamic, measuring C_{TSM} values at a higher spatiotemporal level could assist in preserving the aquatic ecology and shed light on the driving force of TSM. As field measurements can be costly and have limited spatiotemporal scope, using satellite remote sensing to assess C_{TSM} provides a more feasible alternative (Merina, 2017). Sediment accumulation in the Hugli estuary is concentrated downwards due to decreased water flow velocity. Continuous evaluation of C_{TSM} is required for the operation of the Haldia port situated downstream, which will further help in the timely maintenance of siltation and dredging for an extended period of time (Prasad & Singh, 2010). Kumar developed a regional algorithm to monitor and retrieve C_{TSM} via in situ C_{TSM} revival and in situ reflectance data from the Sentinel 2 MSI satellite data on the Hugli estuary. In situ assessments were performed using the Satlantic HyperOCR ("Hyperspectral Ocean Color Radiometer"), and water sampling was acquired at 0.5 m depth. Through these techniques, they estimated that the C_{TSM} range in the Hugli estuary varies between 132 and 540 mg/L during the pre-monsoon period in May (Premkumar et al., 2021a). Another investigation by Jayaram et al. (2021), seasonal and interannual variation in TSM of the Hugli estuary and adjacent Sundarbans region, was conducted using MODIS time-series dataset over a long period (2003–2018). During this period, TSM in the Hugli estuary was higher in 2004 than in 2005 and higher in 2012 and 2013 than in 2015. During the southwest monsoon season, TSM concentrations in the Hugli estuary were the highest (>250 g m^{-3}) due to increased river flow, compared to the northeast monsoons and inter-monsoon (100–150 g m^{-3}) (Jayaram et al., 2021).

3.1.2 CDOM

"Colored or chromophobic dissolved organic matter" (CDOM) is a type of dissolved organic matter that absorbs short wavelength radiation that imparts color to natural waters (Coble, 2007). CDOM can be found in marine and freshwater ecosystems (Kirk, 1994). CDOM being a subset or component of the overall DOM in the hydrosphere which absorbs light across UV range (UV-A: 315–400 nm; UV-B: 280–315 nm) and visible range (400–750 nm). CDOM comprises a mix of aromatic and aliphatic polymers derived from aquatic and floral degradation (Moran et al., 1999; Raymond & Bauer, 2000). CDOM is optically detectable and significantly impacts the global carbon cycle. In addition to flora degradation, CDOM is caused by human operations such as logging, cultivation, sewage discharge, wetlands drainage, and debris transport across freshwater and estuary systems (Coble, 1996) (Fig. 7.3).

As CDOM levels are higher in coastal and estuarine areas than offshore regions, detecting the sinks and sources of CDOM is crucial to enhance our understanding of its contribution to biogeochemical processes, such as the carbon cycle (Vodacek et al., 1995).

The Hugli estuary, which is a freshwater-seawater mixing, has studies conducted by Das et al. (2016), 2017a, 2017b) to analyze the variability of CDOM from October 2014 to January 2016 across nine locations. It was found during 2014–2016 that the absorbance of CDOM at 440 nm, i.e., aCDOM (440), range is 0.1002–0.6631 m^{-1}. The average amount of aCDOM (440) varied significantly across the three seasons. During the pre-monsoon season, aCDOM (440) range is 0.1200 ± 0.0327 m^{-1}. In monsoons, aCDOM (440) range is 0.3064 ± 0.1595 m^{-1}, and in post-monsoons aCDOM (440) range is 0.1621 ± 0.0790 m^{-1}. Furthermore, the spatial distribution analysis showed that aCDOM (440) was higher nearby confluence and gradually decreased offshore.

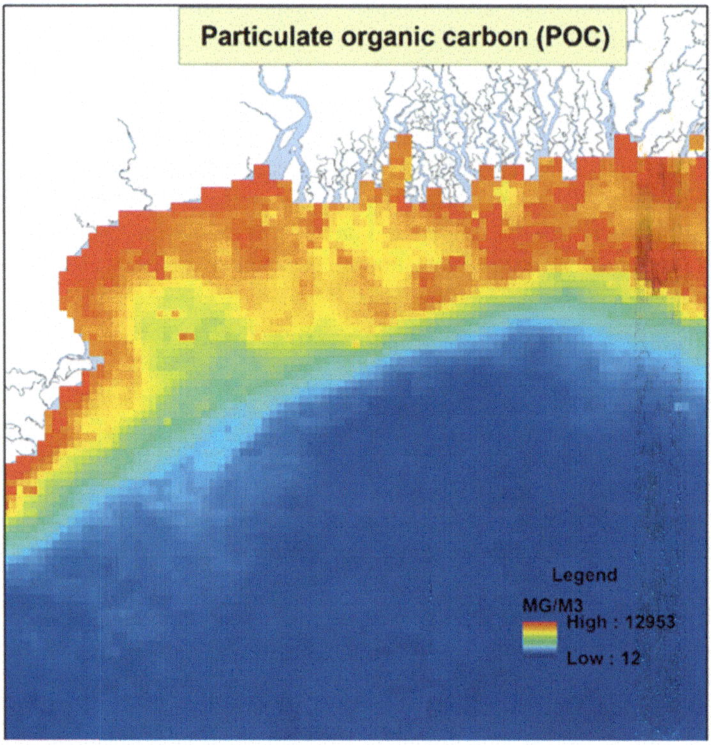

Fig. 7.3 POC distribution along the northern Bay of Bengal

3.1.3 Chlorophyll

In estuaries, chlorophyll constitutes an essential bio-optical attribute because it indicates the biomass of phytoplankton, which influences the optical properties of the water (Goela et al., 2015). The chlorophyll content in water is an accurate indicator of biotic production within all aquatic ecosystems (Mondal et al., 2024a). The Chl-a pigment present in phytoplankton is a key indicator for biomass which influences light absorption and scattering in water column. The characteristic absorption of Chl-a in red and blue wavelength and scattering in green wavelength impact water color and clarity (Dierssen et al., 2006) and is used for biomass estimation (Cloern, 1996). The Chl-a concentration in estuarine water is proportional to primary production which lays the foundation of aquatic food chain that sustains marine species of higher trophic levels (Cloern et al., 2014). The chlorophyll concentration in the Hugli estuary typically shows a seasonal trend with high levels of Chl-a and phytoplankton bloom during monsoon as a result of massive freshwater and nutrient input, with additional tidal mixing (Portel & Rama Chandra Prasad, 2024).

Recent field sampling (Mondal et al., 2024a) at Fraserganj and Sagar Island in the Hugli estuary based on the technique developed by Strickland and Parsons

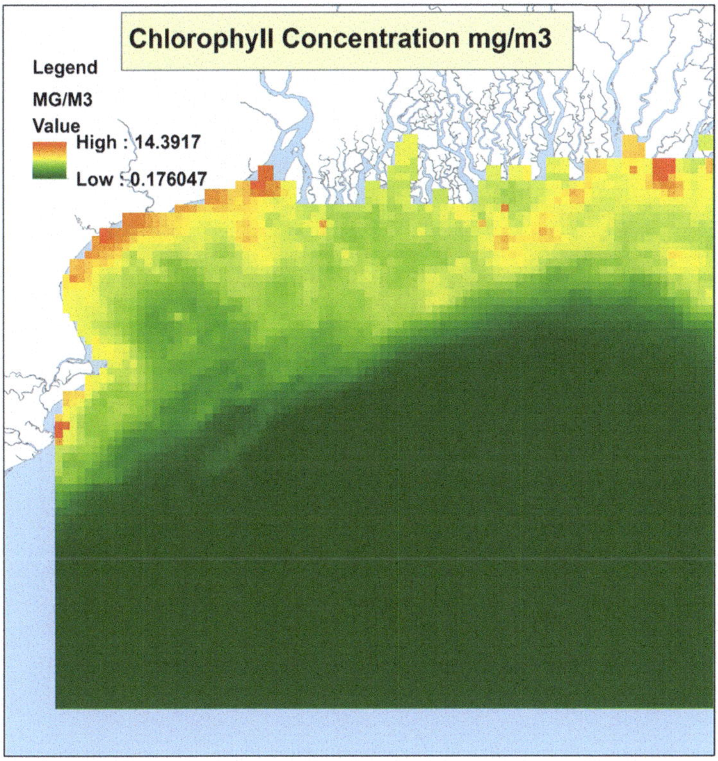

Fig. 7.4 Chlorophyll concentration in the Northern Bay of Bengal

(1972) showed a good correlation between in situ measurements (3.03 mg m^{-3}) and satellite-derived Chl-a concentration (2.96 mg m^{-3}). Maximum concentration of Chl-a was observed during post-monsoon (3.17 mg m^{-3}) suggest that the contribution of continental shelf towards phytoplankton production is greater than continental slope. This observation denotes that fishing activities should be more productive along continental shelf in coastal BoB (Mondal et al., 2024b) (Fig. 7.4).

3.1.4 Temperature

Temperature is one of the key bio-optical property which influence several physical, chemical, and biological processes within an estuary. Therefore, proper management and maintenance of an estuary involves comprehending the relationship between temperature and other bio-optical properties (Dias et al., 2020). Alteration of temperature leads to change in viscosity and density of water which in turn impact light absorption and scattering. Higher temperature in water column leads to reduced density and viscosity, resulting in decreased scattering from suspended particles (Kirk, 1994). Temperature also regulates the density of estuarine water,

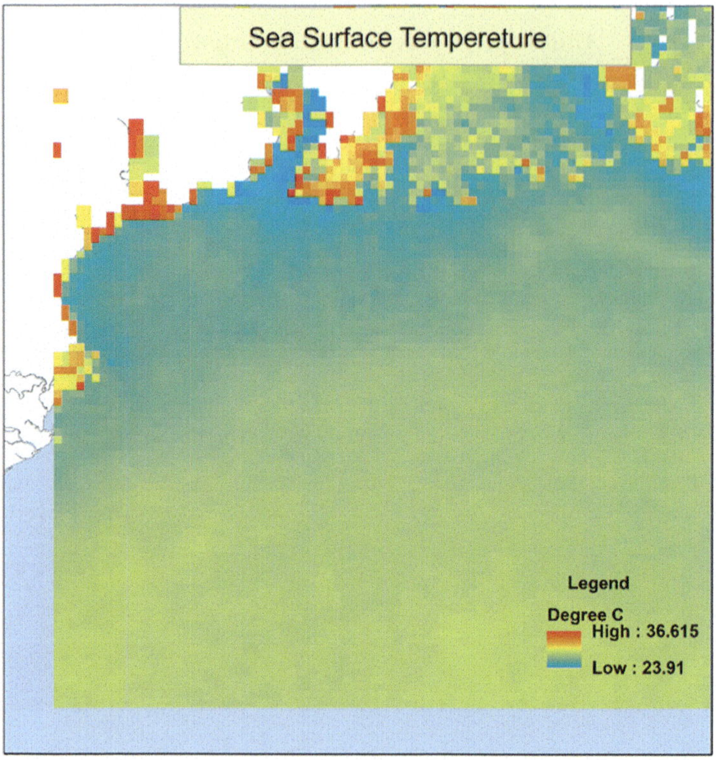

Fig. 7.5 Average sea surface temperature in the Northern Bay of Bengal

effectively regulating mixing and circulation by upwelling and downwelling which affects the distribution of organic matter and suspended sediments (Dutta et al., 2019). Sadhuram et al. (2005) found that the water temperature of the Hugli estuary varies by 10 °C between winter (30–31 °C) and summer (20–21 °C). However, vertical temperature in the Hugli estuary did not fluctuate significantly since the estuary is well-mixed. Its water density ranged between 999.5 and 1000.4 kg m^{-3} during summer under flood and ebb tides. During post-monsoons in the Hugli estuary, temperatures range between 27 and 28 °C, with density values below 1000 kg m^{-3} under flood and ebb tides (Fig. 7.5).

3.1.5 Salinity

Salinity, which refers to the concentration of salts dissolved in water, is essential in estuarine ecosystems. It affects the water's optical characteristics and, in turn, could be calculated by optical techniques (Chong et al., 2014). Monitoring water quality, managing the environment, and conducting ecological research depends on understanding salinity trends and how they affect estuarine ecosystems (Montagna et al., 2012).

Under intense tidal action, estuarine waters result in significant variations in salinity levels among flood and ebb tides within a single day (Basu et al., 1970). During the dry season, saline intrusion in the Hugli estuary stretches 70 km from the mouth. However, this varies periodically. Salinity was recorded at eight locations using the SBE Seacat profiler during flood and ebb tides by Sadhuram et al. (2005), and it varied both spatio-temporally and seasonally. During the summer, the salinity ranged from 4.5 to 6.0 ppt at ebb tides and 7.0 to 13.0 ppt (parts per thousand) during flood tides. Due to limited flow, salinity in the post-monsoon period dropped dramatically to 0.2–1.2 ppt at ebb tides and 1.8–2.4 ppt at flood tides. During the flood tide, two stations had higher bottom salinity (2.6 ppt). In a recent study on salinity, Mondal used the argentometric titration technique (Strickland & Parsons, 1972) to assess salinity in the Hugli estuary. The average salinities recorded in the Hugli estuary were 1.43 ± 0.66 in pre-monsoon, 0.78 ± 0.60 in monsoon, and 1.18 ± 0.45 in post-monsoon.

Salinity of the Hugli estuary has also been influenced by the construction of Farakka barrage in upstream. Before the barrage's construction, the freshwater zone was relatively small, spanning about 110 km between Nabadwip–Nawabganj–Barrackpore due to limited freshwater discharge (Sharma et al., 2013). David (1954) recorded a salinity range of 0.004–0.20 "parts per thousand," i.e., ppt (sometimes reaching to 1.16 ppt) at Barrackpore during this period. However, the introduction of increased freshwater inflow from the Farakka barrage caused a significant shift in the salinity patterns and range of the Hugli Estuary. The salinity range observed by David (1954) at the Barrackpore–Nawabganj region is now found beyond the Godakhali–Uluberia region, approximately 60 km downstream. Nandy et al. (1983) underscored the necessity of redefining the zones in the Hooghly Estuary, recognizing the crucial role of the Farakka barrage discharge in this process. Salinity is also key regulator of fish and other aquatic organisms' distribution in the Hugli Estuary (Sharma et al., 2013). Increased freshwater intake from the Farakka barrage has altered the salinity regime in the Hooghly estuary, affecting the distribution and quantity of aquatic organisms such as plankton. Nandy et al. (1983) showed that formation of the Farakka barrage reduced the salinity in the Hugli estuary resulting in three to five times higher plankton population.

3.1.6 Dissolved Oxygen

The aerobic communities in the marine ecosystem are dependent on dissolved oxygen (DO) which is crucial in trophodynamics and can act as an indicator for water quality (Hull et al., 2000). Several factors can influence DO levels in estuarine water like water temperature, atmospheric gas pressure, salinity, and various biological processes (Reid, 1961). The processes that maintain DO balance in estuary include photosynthesis, aerobic respiration, sediment oxygen demand, chemical oxidation, and other hydrodynamic processes (Lunardini & Di Cola, 2000). The aquatic species in the Hugli estuary are presently at risk as the continuous mangrove degradation is going to reduce the available DO in water which is essential for the marine species to thrive (Mandal et al., 2012).

Numerous experimental studies have examined the dynamics of DO across the Hugli estuary (Bhunia, 1979; Sarkar et al., 1986; Mukhopadhyay et al., 2006; Mandal et al., 2012; Sharma et al., 2013). The modified Winkler's-Iodometric technique (Strickland & Parsons, 1972) is used by Mandal et al. (2012) for measuring DO in the Hugli estuary water. Their field estimations in the Hugli estuary showed that dissolved oxygen (Do) levels were higher post-monsoon (5.22 ± 0.30 mg/L), peaking in February (5.57 mg/L) and lower pre-monsoon (4.59 ± 0.38 mg/L), reaching a trough in May (4.43 mg/L). There is an inverse relationship between DO dynamics and water temperature, with higher values observed during pre-monsoon (29.5–31 °C) and lower values observed during winter (20.5–22.5 °C) (Mandal et al., 2012). The study also indicates an inverse relationship between DO and salinity only before the monsoon, with no clear correlation during the other seasons. The salinity of the creek water is the highest (21.82 mg/L) before the monsoon and lowest (2.04 mg/L) during the monsoon (Mandal et al., 2012).

3.1.7 PFZ

Certain regions of the seas and estuaries are known as "Potential Fishing Zones" (PFZ), which are considered to have a high concentration of fish. Salinity, SST, Chl-a, and turbidity are examples of bio-optical water properties used to define these zones (Daqamseh et al., 2019). Based on these properties derived from satellite datasets (MODIS, OCM, NOAA AVHRR, and SeaWiFS), PFZs can be delineated (Nayak et al., 2023; Solanki et al., 2003). It is essential to comprehend PFZ about the Hugli estuary to carry out sustainable fishery management and increase the output of nearby fishing communities. A study by Dutta et al. (2016) aimed to predict Potential Fishery Zones (PFZs) in the Northern Bay of Bengal, which is adjacent to the Hugli estuary. The study found a positive correlation ($R = 0.748$, $P < 0.0001$) between "Catch Per Unit Effort" (CPUE), phytoplankton biomass, and essential nutrients such as dissolved inorganic phosphate, nitrate, and silicate. According to this research, high-nutrient water stimulates the growth of phytoplankton biomass, which leads to fish aggregation and can result in the formation of PFZs (Table 7.3).

The bio-optical characteristics of estuarine waters are different globally due to differences in geochemical, hydrological, and ecological conditions. The Hugli Estuary, a branch of the Ganges-Brahmaputra delta, has significantly different bio-optical properties compared to other estuaries, mainly due to high sediment concentration, organic matter from the adjacent Sundarbans mangrove forest, and anthropogenic impacts. A characteristic feature of the Hugli Estuary is the high turbidity due to sediment supply and tidal stirring. This is dissimilar with other estuaries such as the Chesapeake Bay in the United States where turbidity changes with the season due to phytoplankton growth. Low Secchi depth of the Hugli Estuary because of high turbidity is a major factor of low light penetration, thus low

Table 7.3 The comparison between bio-optical properties of Hugli water and other estuaries worldwide

Parameter	The Hugli Estuary (Mondal et al., 2024b; Premkumar et al., 2021b)	Amazon River Estuary (Cesar et al., 2023; Valerio et al., 2018)	Chesapeake Bay (McKibben et al., 2024)	Pearl River Estuary (Deng et al., 2024; Jiang et al., 2024)	Thames Estuary (Bailey et al., 2024; Defontaine et al., 2024)	General Trend (Li et al., 2024)
CDOM (colored dissolved organic matter)	High due to high organic load from the Sundarbans and human activities	Moderate, influenced by seasonal rainfall	Low to moderate, varies with seasonal algae blooms	High, influenced by industrial pollution	Moderate, due to urban runoff	Increases in areas with high organic and nutrient load
Turbidity	High due to suspended sediments and tidal forces	Moderate to high, seasonal variation	Moderate	High, especially near urban areas	Moderate	Typically high in river-fed estuaries
Chlorophyll-a concentration	Low, nutrient limitations inhibit phytoplankton growth	High, nutrient-rich environment	Moderate to high	High, nutrient-rich due to pollution	Low to moderate	Higher in nutrient-rich waters
Secchi depth	Shallow due to high turbidity	Moderate	Moderate	Shallow, affected by pollution	Moderate	Lower in turbid estuaries
Backscatter (bbp)	High, reflecting high sediment load	Moderate	Moderate	High, due to high particle concentration	Low to moderate	Varies based on sediment and particle load

primary productivity as evident from low chlorophyll-a concentrations as compared to other estuaries like the Amazon River estuary that has high phytoplankton density due to high nutrient availability. Another important characteristic is the concentration of Colored Dissolved Organic Matter, which is rather high in the waters of the Hugli. CDOM concentrations in estuarine systems are normally associated with the organic inputs from upstream and local vegetation, and higher CDOM levels are recorded where there are high inputs of organic materials. The Hugli's higher CDOM has been attributed with the Sundarbans' vegetation and anthropogenic effects, whereas moderate CDOM of the Thames is due to urban water discharge. Other backscattering properties affecting the light absorption and scattering within the water column are also observed in the Hugli Estuary due to high suspended particle loads. This is a departure from more developed estuaries like the Thames and Pearl River estuaries where backscattering is affected by industrial effluents.

However, nutrient-concentrated estuaries such as the Pearl River in China are found to have higher chlorophyll-a values, which indicates higher phytoplankton productivity, which is characteristic of nutrient-enriched waters. However, the nutrient concentrations in the Hugli Estuary are not as high to support such phytoplankton production because of interaction between organic matter and turbidity. Nevertheless, the bio-optical signatures of the Hugli Estuary can be used to understand the impact of natural vegetation and anthropogenic activities on the water quality. Comparative analysis with other estuaries reveals the specificity of the ecological effects in the Hugli, thus pointing to the need for context-based approaches to the management of the natural environment. Recognizing such differences will help to determine the state and sustainability of estuarine ecosystems globally in regard to the processes of urbanization and climate change.

4 Conclusion

This paper presents a comprehensive overview of the bio-optical properties of the Hugli estuary and northern BoB. Estuaries are the main conduits for the transfer of substantial volumes of dissolved and particulate matter from terrestrial habitats into the ocean. They are also the main connector between land and ocean systems. The state of West Bengal receives most of its freshwater from the Hugli River; therefore, it is crucial to monitor the Hugli River estuary and its pollution zone for management purposes. The high population pressure and associated economic growth activities including transportation, agriculture, industry, and recreation in the coastal West Bengal has put the Hugli estuary at risk.

Water quality study conducted in the field is usually costly and time-consuming. On the contrary, satellite-based retrieval of water quality parameters for monitoring the aquatic environment is effective. It is difficult to monitor water quality using remote sensing in the estuaries because they belong to the optically complex case II waters, where the optical properties depend not only on the phytoplankton but also on sediments and dissolved organic matter. In comparison to open marine habitats, estuarine waters are also distinguished by comparatively higher quantities of suspended sediment, phytoplankton, and CDOM. Therefore, a brief description of light interaction in the saline estuarine water has been presented in this paper with special emphasis on sediments, CDOM, chlorophyll, temperature, dissolved oxygen, and salinity. This review of the bio-optical parameters in the Hugli estuary will be helpful in future research and developing a monitoring protocol. The parameters estimated in this study could be used for water quality measurements and instigating policies by the governmental bodies for protecting the major freshwater source in West Bengal.

References

Abdelly, C., Flowers, T., Gul, B., Koyro, H.-W., & Savouré, A. (2022). Recent advances in research on halophytes: From fundamental to applied aspects. *Environmental and Experimental Botany, 193*, 104683. https://doi.org/10.1016/j.envexpbot.2021.104683

Abu Shmeis, R. M. (2018). *Water chemistry and microbiology. Volume 81: Comprehensive analytical chemistry*. Elsevier BV.

Aghaleh, M., Niknam, V., Ebrahimzadeh, H., & Razavi, K. (2011). Effect of salt stress on physiological and antioxidative responses in two species of Salicornia (S. Persica and S. Europaea). *Acta Physiologiae Plantarum, 33*(4), 1261–1270. https://doi.org/10.1007/s11738-010-0656-x

Akhand, A., Chanda, A., Manna, S., Das, S., Hazra, S., Roy, R., Choudhury, S. B., Rao, K. H., Dadhwal, V. K., Chakraborty, K., Mostofa, K. M. G., Tokoro, T., Kuwae, T., & Wanninkhof, R. (2016). A comparison of CO2 dynamics and air-water fluxes in a river-dominated estuary and a mangrove-dominated marine estuary. *Geophysical Research Letters, 43*(22), 711–726. https://doi.org/10.1002/2016GL070716

Alrifai, O., Hao, X., Marcone, M. F., & Tsao, R. (2019). Current review of the modulatory effects of LED lights on photosynthesis of secondary metabolites and future perspectives of microgreen vegetables. *Journal of Agricultural and Food Chemistry, 67*(22), 6075–6090. https://doi.org/10.1021/acs.jafc.9b00819

Annunziata, M. G., Apelt, F., Carillo, P., Krause, U., Feil, R., Mengin, V., Lauxmann, M. A., Köhl, K., Nikoloski, Z., & Stitt, M. (2017). Getting back to nature: A reality check for experiments in controlled environments. *Journal of Experimental Botany, 68*(16), 4463–4477.

Araus, J. L., Rezzouk, F. Z., Thushar, S., Shahid, M., Elouafi, I. A., Bort, J., & Serret, M. D. (2021). Effect of irrigation salinity and ecotype on the growth, physiological indicators and seed yield and quality of Salicornia europaea. *Plant Science, 304*, 110819. https://doi.org/10.1016/j.plantsci.2021.110819

Arora, C., & Bhaskaran, P. K. (2013). Numerical modeling of suspended sediment concentration and its validation for the Hooghly estuary, India. *Coastal Engineering Journal, 55*(2), 1350006-1-1350006–1350023. https://doi.org/10.1142/S057856341350006X

Asseng, S., Guarin, J. R., Raman, M., Monje, O., Kiss, G., Despommier, D. D., Meggers, F. M., & Gauthier, P. P. G. (2020). Wheat yield potential in controlled-environment vertical farms. *Proceedings of the National Academy of Sciences, 117*(32), 19131–19135. https://doi.org/10.1073/pnas.2002655117

Atkinson, M. J., & Bingman, C. (1997). Elemental composition of commercial seasalts. *Journal of Aquariculture and Aquatic Sciences, 8*(2), 39–43.

Bailey, L. P., Dorrell, R. M., Kostakis, I., McKee, D., Parsons, D., Rees, J., Strong, J., Simmons, S., & Forster, R. (2024). Monopile-induced turbulence and sediment redistribution form visible wakes in offshore wind farms. *Frontiers in Earth Science, 12*. https://doi.org/10.3389/feart.2024.1383726

Bagchi, K. (1944): The Ganges Delta. *Calcutta University*, Calcutta. 120 p
Bandyopadhyay, S. (2000). Coastal changes in the perspective of long term evolution of an estuary: Hugli, West Bengal, India. In *International Quaternary Seminar on INQUA Shoreline Indian Ocean Sub-Commission* (pp. 103–115).
Basu, A. K., Ghosh, B. B., & Pal, R. N. (1970). Comparison of the polluted Hooghly estuary with the unpolluted Matlah estuary, India. *Journal (Water Pollution Control Federation), 42*(10), 1771–1781. http://www.jstor.org/stable/25036797
Basu, S. R., & Chakraborty, S. (1972). Some considerations over the decay of the Bhagirathi drainage system. In *The Bhagirathi-Hooghly Basin, Proceedings of Interdisciplinary Symposium* (pp. 57–77).
Ben Hassine, A., Ghanem, M. E., Bouzid, S., & Lutts, S. (2009). Abscisic acid has contrasting effects on salt excretion and polyamine concentrations of an inland and a coastal population of the Mediterranean xero-halophyte species Atriplex halimus. *Annals of Botany, 104*(5), 925–936.
Bhaumik, U., & Sharma, A. P. (2012). *Present status of Hilsa in Hooghly-Bhagirathi River*. Retrieved from http://www.cifri.res.in/Bulletins/Bulletin%20No.179.pdf
Bhunia, A. B. (1979). *Ecology of the tidal creeks mudflats of Sagar Island sunderbans West Bengal*.
Biswas, H., Mukhopadhyay, S. K., De, T. K., Sen, S., & Jana, T. K. (2004). Biogenic controls on the air—Water carbon dioxide exchange in the Sundarban mangrove environment, northeast coast of bay of Bengal, India. *Limnology and Oceanography, 49*(1), 95–101. https://doi.org/10.4319/lo.2004.49.1.0095
Biswas, H., Mukhopadhyay, S. K., Sen, S., & Jana, T. K. (2007). Spatial and temporal patterns of methane dynamics in the tropical mangrove dominated estuary, NE coast of Bay of Bengal, India. *Journal of Marine Systems, 68*(1), 55–64. https://doi.org/10.1016/j.jmarsys.2006.11.001
Breckle, S.-W. (2002). Chapter 3. Salinity. In *Halophytes and salt affected natural ecosystems*.
Bricaud, A., & Stramski, D. (1990). Spectral absorption coefficients of living phytoplankton and nonalgal biogenous matter: A comparison between the Peru upwelling area and the Sargasso Sea. *Limnology and Oceanography, 35*(3), 562–582.
Brudvig, G. W., Beck, W. F., & Paula, J. C. (1989). Mechanism of photosynthetic water oxidation. *Annual Review of Biophysics and Biophysical Chemistry, 18*(1), 25–46.
Bukata, R. P., Jerome, J. H., Kondratyev, A. S., & Pozdnyakov, D. V. (2018). *Optical properties and remote sensing of inland and coastal waters*. CRC Press.
Cesar, G. M., Rudorff, N. M., Oliveira, A. L., Valerio, A. M., Chuqui, M. G., Pompeu, M., Gaeta, S. A., Frouin, R., & Kampel, M. (2023). Bio-optical properties of South Brazil bight coastal waters and implications for satellite chlorophyll-a concentration retrieval. *International Journal of Remote Sensing, 44*(7), 2428–2457. https://doi.org/10.1080/01431161.2023.2201385
Chong, Y. J., Khan, A., Scheelbeek, P., Butler, A., Bowers, D., & Vineis, P. (2014). Climate change and salinity in drinking water as a global problem: Using remote-sensing methods to monitor surface water salinity. *International Journal of Remote Sensing, 35*(4), 1585–1599.
Cloern, J. E. (1996). Phytoplankton bloom dynamics in coastal ecosystems: A review with some general lessons from sustained investigation of San Francisco Bay, California. *Reviews of Geophysics, 34*(2), 127–168. https://doi.org/10.1029/96RG00986
Cloern, J. E., Foster, S. Q., & Kleckner, A. E. (2014). Phytoplankton primary production in the world's estuarine-coastal ecosystems. *Biogeosciences, 11*(9), 2477–2501.
Coble, P. G. (1996). Characterization of marine and terrestrial DOM in seawater using excitation-emission matrix spectroscopy. *Marine Chemistry, 51*(4), 325–346. https://doi.org/10.1016/0304-4203(95)00062-3
Coble, P. G. (2007). Marine optical biogeochemistry: The chemistry of ocean color. *Chemical Reviews, 107*(2), 402–418. https://doi.org/10.1021/cr050350+
Daqamseh, S. T., Al-Fugara, A., Pradhan, B., Al-Oraiqat, A., & Habib, M. (2019). MODIS derived sea surface salinity, temperature, and chlorophyll-a data for potential fish zone mapping: West red sea coastal areas, Saudi Arabia. *Sensors, 19*(9), 2069.

Das, S., Das, I., Giri, S., Chanda, A., Maity, S., Lotliker, A. A., Kumar, T. S., Akhand, A., & Hazra, S. (2017a). Chromophoric dissolved organic matter (CDOM) variability over the continental shelf of the northern bay of Bengal. *Oceanologia, 59*(3), 271–282. https://doi.org/10.1016/j.oceano.2017.03.002

Das, S., Hazra, S., Giri, S., Das, I., Chanda, A., Akhand, A., & Maity, S. (2017b). *Light absorption characteristics of chromophoric dissolved organic matter (CDOM) in the coastal waters of northern Bay of Bengal during winter season.*

Das, S., Hazra, S., Lotlikar, A. A., Das, I., Giri, S., Chanda, A., Akhand, A., Maity, S., & Kumar, T. S. (2016). Delineating the relationship between chromophoric dissolved organic matter (CDOM) variability and biogeochemical parameters in a shallow continental shelf. *Egyptian Journal of Aquatic Research, 42*(3), 241–248. https://doi.org/10.1016/j.ejar.2016.08.001

DataStream Initiative. (2021). *A water monitor's guide to water quality.* Retrieved from https://datastreamorg.sharepoint.com/sites/Datastream/Shared%20Documents/Forms/AllItems.aspx?id=%2Fsites%2FDatastream%2FShared%20Documents%2FFiles%2F6%20%2D%20Outreach%20and%20User%20Related%2FScience%20Explainers%2F5%20%2D%20Printable%20PDF%2FEnglish%2FChapters%2FPhysical%5FChemical%5FChapter%2Epdf&parent=%2Fsites%2FDatastream%2FShared%20Documents%2FFiles%2F6%20%2D%20Outreach%20and%20User%20Related%2FScience%20Explainers%2F5%20%2D%20Printable%20PDF%2FEnglish%2FChapters&p=true&ga=1

David, A. (1954). A preliminary survey of the fish and fisheries of a five-mile stretch of the Hooghly River near Barrackpore. *Indian Journal of Fisheries, 1*(1 & 2), 231–255.

Davis, M. L., & Cornwell, D. A. (1991). Introduction to environmental engineering.

De, T. K., De, M., Das, S., Chowdhury, C., Ray, R., & Jana, T. K. (2011). Phytoplankton abundance in relation to cultural eutrophication at the land-ocean boundary of Sunderbans, NE coast of bay of Bengal, India. *Journal of Environmental Studies and Sciences, 1*(3), 169–180. https://doi.org/10.1007/s13412-011-0022-3

De, T. K., Mukherjee, A., Das, S., De, M., & Maiti, T. K. (2015). Interrelationship between planktonic diatoms and selected governing physicochemical parameters of the Hooghly estuary, Bay of Bengal. *International Journal of Marine Science, 5*(47).

Defontaine, S., Jalón-Rojas, I., Sottolichio, A., Gratiot, N., Legout, C., & Lienart, C. (2024). Factors controlling mud floc settling velocity in a highly turbid macrotidal fluvial-estuarine system. *Journal of Geophysical Research: Oceans, 129*(10), e2024JC021558. https://doi.org/10.1029/2024JC021558

Deng, L., Hu, S., Chen, S., Zeng, X., Wang, Z., & Liu, S. (2024). Particle backscattering efficiency as an indicator of suspended particle composition and size in the Pearl River estuary. *Marine Pollution Bulletin, 209*, 117189. https://doi.org/10.1016/J.MARPOLBUL.2024.117189

Dias, A., Kurian, S., & Thayapurath, S. (2020). Influence of environmental parameters on bio-optical characteristics of colored dissolved organic matter in a complex tropical coastal and estuarine region. *Estuarine, Coastal and Shelf Science, 242*, 106864. https://doi.org/10.1016/j.ecss.2020.106864

Dierssen, H. M., Kudela, R. M., Ryan, J. P., & Zimmerman, R. C. (2006). Red and black tides: Quantitative analysis of water-leaving radiance and perceived color for phytoplankton, colored dissolved organic matter, and suspended sediments. *Limnology and Oceanography, 51*(6), 2646–2659. https://doi.org/10.4319/lo.2006.51.6.2646

Dutta, M. K., Kumar, S., Mukherjee, R., Sanyal, P., & Mukhopadhyay, S. K. (2019). The post-monsoon carbon biogeochemistry of the Hooghly–Sundarbans estuarine system under different levels of anthropogenic impacts. *Biogeosciences, 16*(2), 289–307.

Dutta, S., Chanda, A., Akhand, A., & Hazra, S. (2016). Correlation of phytoplankton biomass (chlorophyll-a) and nutrients with the catch per unit effort in the PFZ forecast areas of Northern Bay of Bengal during simultaneous validation of winter fishing season. *Turkish Journal of Fisheries and Aquatic Sciences, 16*(4), 767–777.

Ebbing, D. D., & Gammon, S. D. (1996). General chemistry. *Energy, 2*(2), 2–13.

Wenner, E. L. (2001). The National Estuarine Research Reserves Program to monitor and preserve estuarine waters. *Coastal Management, 29*(1), 1–17. https://doi.org/10.1080/089207501750057301

EPA. (2006). Chapter 11: pH and Alkalinity. In *Volunteer Estuary Monitoring Manual. A methods manual* (2nd ed., pp. 1–13). United States Environmental Protection Agency. Retrieved from http://www.epa.gov/owow/estuaries/monitor/

Estuary Watch. (2011). *Monitoring your estuary. A methods manual for communities* (pp. 1–16). Estuary Watch. Retrieved from https://www.estuarywatch.org.au/resources/009_ew_cemm_manual_physico_chem.pdf

Fox, M. (2010). *Optical properties of solids* (Vol. 3). Oxford University Press.

Gensemer, R. W., & Playle, R. C. (1999). The bioavailability and toxicity of aluminum in aquatic environments. *Critical Reviews in Environmental Science and Technology, 29*(4), 315–450. https://doi.org/10.1080/10643389991259245

Giardino, C., Brando, V. E., Gege, P., Pinnel, N., Hochberg, E., Knaeps, E., Reusen, I., Doerffer, R., Bresciani, M., Braga, F., Foerster, S., Champollion, N., & Dekker, A. (2019). Imaging spectrometry of inland and coastal waters: State of the art, achievements and perspectives. *Surveys In Geophysics, 40*(3), 401–429. https://doi.org/10.1007/s10712-018-9476-0

Goela, P. C., Icely, J., Cristina, S., Danchenko, S., Angel DelValls, T., & Newton, A. (2015). Using bio-optical parameters as a tool for detecting changes in the phytoplankton community (SW Portugal). *Estuarine, Coastal and Shelf Science, 167*, 125–137. https://doi.org/10.1016/j.ecss.2015.07.037

Haltrin, V. (2006). Absorption and scattering of light in natural waters. In *Light scattering reviews single and multiple light scattering* (pp. 445–486). https://doi.org/10.1007/3-540-37672-0_10

Hanrahan, G. (2012). Chapter 3. Aqueous chemistry. In G. Hanrahan (Ed.), *Key concepts in environmental chemistry* (pp. 73–106). Academic Press. https://doi.org/10.1016/B978-0-12-374993-2.10003-2

Hull, V., Mocenni, C., Falcucci, M., & Marchettini, N. (2000). A trophodynamic model for the lagoon of Fogliano (Italy) with ecological dependent modifying parameters. *Ecological Modelling, 134*(2), 153–167. https://doi.org/10.1016/S0304-3800(00)00358-6

Jayaram, C., Patidar, G., Swain, D., Chowdary, V. M., & Bandyopadhyay, S. (2021). Total suspended matter distribution in the Hooghly River estuary and the Sundarbans: A remote sensing approach. *IEEE Journal of Selected Topics in Applied Earth Observations and Remote Sensing, 14*, 9064–9070. https://doi.org/10.1109/JSTARS.2021.3076104

Jiang, J., Chen, Z., Sun, K., Chen, Y., Mu, M., Sun, Q., & Zeng, S. (2024). Monitoring 36-year water clarity dynamics in turbid waters of the Pearl River estuary, China, using Landsat data. *Ecological Informatics, 81*, 102600. https://doi.org/10.1016/J.ECOINF.2024.102600

Kirk, J. T. O. (1994). *Light and photosynthesis in aquatic ecosystems*. Cambridge University Press.

Kratzer, S., Kowalczuk, P., & Sagan, S. (2017). Bio-optical water quality assessment. In *Biological Oceanography of the Baltic Sea* (pp. 527–545).

Li, J., Antoine, D., & Huot, Y. (2024). Bio-optical variability of particulate matter in the Southern Ocean. *Frontiers in Marine Science, 11*. https://doi.org/10.3389/fmars.2024.1466037

Lide, D. R. (2004). *CRC handbook of chemistry and physics* (Vol. 85). CRC Press.

Lunardini, F., & Di Cola, G. (2000). Oxygen dynamics in coastal and lagoon ecosystems. *Mathematical and Computer Modelling, 31*(4), 135–141. https://doi.org/10.1016/S0895-7177(00)00031-5

Manahan, S. E. (2022). *Environmental chemistry*. CRC Press.

Mandal, S., Debnath, M., Ray, S., Ghosh, P. B., Roy, M., & Ray, S. (2012). Dynamic modelling of dissolved oxygen in the creeks of Sagar island, Hooghly–Matla estuarine system, West Bengal, India. *Applied Mathematical Modelling, 36*(12), 5952–5963. https://doi.org/10.1016/j.apm.2011.10.013

McDowell, D. M. (1995). The development of the river Hooghly for navigation. *Commemoration Volume Port of Calcutta, 125*, 69.

McLusky, D. S. (1989). The estuarine environment. In D. S. McLusky (Ed.), *The estuarine ecosystem* (pp. 1–48). Springer. https://doi.org/10.1007/978-94-011-7616-3_1

Merina, R. N. (2017). Evaluation of sediment using GIS approach in southern Tamilnadu, India. *The Indian Ecological Society, 44*(2), 179–184.

Miller, R. L., Del Castillo, C. E., & McKee, B. A. (2005). *Remote sensing of coastal aquatic environments* (Vol. 511). Springer.

Mitchell, B. G., Bricaud, A., Carder, K., Cleveland, J., Ferrari, G., Gould, R., Kahru, M., Kishino, M., Maske, H., Moisan, T., Moore, L., Nelson, N., Phinney, D., Reynolds, R., Sosik, H., Stramski, D., Tassan, S., Trees, C., Weidemann, A., & Vodacek, A. (2000). Determination of spectral absorption coefficients of particles, dissolved material and phytoplankton for discrete water samples. In *NASA Technical Memorandum* (pp. 125–153).

McKibben, S. M., Schollaert Uz, S., & Palacios, S. L. (2024). Testing a hyperspectral, bio-optical approach to identification of phytoplankton community composition in the Chesapeake Bay estuary. *Earth and Space Science, 11*(5), e2023EA003244. https://doi.org/10.1029/2023EA003244

Mobley, C. D. (1995). The optical properties of water. *Handbook of Optics, 1*(43), 43.

Mondal, I., De, A., Nandi, S., Thakur, S., Raman, M., Jose, F., & De, T. K. (2024a). Estimation of chlorophyll-a, TSM and salinity in mangrove dominated tropical estuarine areas of Hooghly River, North East Coast of bay of Bengal, India using sentinel-3 data. *Acta Geophysica, 72*(1), 303–322. https://doi.org/10.1007/s11600-023-01040-5

Mondal, I., Jha, I., Ariful Hossain, S., De, A., Ahmed Altuwaijri, H., Jose, F., Kumar De, T., Lu, Q.-O., & Nguyet Minh, N. (2024b). Variability of bio-optical properties of Sundarbans mangrove estuarine ecosystem using elemental analysis, sentinel 3 OLCI imageries and neural network models. *Advances in Space Research.* https://doi.org/10.1016/J.ASR.2024.10.059

Montagna, P., Palmer, T. A., & Pollack, J. B. (2012). Hydrological changes and estuarine dynamics (8). Springer Science & Business Media.

Moran, M. A., Sheldon, W. M., & Sheldon, J. E. (1999). Biodegradation of riverine dissolved organic carbon in five estuaries of the southeastern United States. *Estuaries, 22*(1), 55–64. https://doi.org/10.2307/1352927

Morgan, J. P., & McIntire, W. G. (1959). Quaternary geology of the Bengal basin, East Pakistan and India. *Geological Society of America Bulletin, 70*(3), 319–342.

Mudryk, R., & Mitchenko, L. (2023). Physical and chemical properties of water. Ecosoft. Retrieved from https://www.ecosoft.com/post/physical-and-chemical-properties-of-water

Mukhopadhyay, A., Acharyya, R., Habel, M., Pal, I., Pramanick, N., Hati, J. P., Sanyal, M. K., & Ghosh, T. (2023). Upstream river erosion vis-a-vis sediments variability in Hugli estuary, India: A geospatial approach. *Water, 15*(7), 1285.

Mukhopadhyay, S. K., Biswas, H., De, T. K., & Jana, T. K. (2006). Fluxes of nutrients from the tropical river Hooghly at the land–ocean boundary of Sundarbans, NE coast of Bay of Bengal, India. *Journal of Marine Systems, 62*(1), 9–21. https://doi.org/10.1016/j.jmarsys.2006.03.004

Nandy, A. C., Bagchi, M. M., & Mazumder, S. K. (1983). Ecological changes in the Hooghly estuary due to water release from Farakka barrage [India]. *Mahasagar, 16.*

Nayak, R. K., Swapna, M., Manche, S. S., Mohanty, P. C., Sheshasai, M. V. R., Dadhwal, V. K., & Kumar, R. (2023). Assessment of chlorophyll-a seasonal cycle in the North Indian Ocean using observations from OCM2, MODIS, and SeaWiFS. *Journal of the Indian Society of Remote Sensing, 51*(2), 229–246. https://doi.org/10.1007/s12524-022-01642-4

Nayak, S. (2004). Role of remote sensing to integrated coastal zone management. In *XXth congress of the International Society for Photogrammetry and Remote Sensing (Istanbul, Turkey), Commission* (Vol. 7, p. 18).

NOAA. (2021). *Classifying estuaries: By water circulation estuaries tutorial.* National Oceanic and Atmospheric Administration (NOAA). Retrieved from https://oceanservice.noaa.gov/education/tutorial_estuaries/est05_circulation.html

Onojake, M. C., Obi, C., & Mahmoud, A. E. D. (2023). Water quality monitoring using sensors and models. In *Artificial intelligence and modeling for water sustainability* (pp. 97–127). CRC Press.

Paesani, F., & Voth, G. A. (2009). The properties of water: Insights from quantum simulations. *The Journal of Physical Chemistry B, 113*(17), 5702–5719. https://doi.org/10.1021/jp810590c

Papadopoulos, M. G., Sadlej, A. J., & Leszczynski, J. (2006). *Non-linear optical properties of matter*. Springer.

Portel, V., & Rama Chandra Prasad, P. (2024). Assessment of water quality parameters in the Hooghly estuary, India, using Sentinel-3 and global biogeochemical analysis and forecasts products. *Thalassas: An International Journal of Marine Sciences*. https://doi.org/10.1007/s41208-024-00678-2

Prasad, A. K., & Singh, R. P. (2010). Chlorophyll, calcite, and suspended sediment concentrations in the bay of Bengal and the Arabian Sea at the river mouths. *Advances in Space Research, 45*(1), 61–69. https://doi.org/10.1016/j.asr.2009.07.027

Premkumar, R., Venkatachalapathy, R., & Visweswaran, S. (2021a). Mapping of total suspended matter based on sentinel-2 data on the Hooghly River, India. *Indian Journal of Ecology, 48*(1), 159–165.

Premkumar, R., Venkatachalapathy, R., & Visweswaran, S. (2021b). Bio-optical studies on chlorophyll-a concentration in Hooghly River, India. *Materials Today: Proceedings, 47*, 488–492. https://doi.org/10.1016/J.MATPR.2021.05.034

Raven, J. A., Gobler, C. J., & Hansen, P. J. (2020). Dynamic CO2 and pH levels in coastal, estuarine, and inland waters: Theoretical and observed effects on harmful algal blooms. *Harmful Algae, 91*, 101594. https://doi.org/10.1016/j.hal.2019.03.012

Raymond, P. A., & Bauer, J. E. (2000). Bacterial consumption of DOC during transport through a temperate estuary. *Aquatic Microbial Ecology, 22*(1), 1–12.

Reaks, H. G. (1919). Report on the physical and hydraulic characteristics of the delta. In *Stevenson Moore et al. (1919)* (pp. 29–132).

Reid, G. K. (1961). Ecology of inland waters and estuaries.

Roy, S. C. (1969). Hydraulic investigations on behalf of Hooghly estuary. In *Metteilungen Des Franzius, Inst Fur Grund-Und-Wasserban Der Technischen Uni Hannover* (p. 149).

Sadhuram, Y., Sarma, V. V., Murthy, T. V. R., & Rao, B. P. (2005). Seasonal variability of physico-chemical characteristics of the Haldia channel of Hooghly estuary, India. *Journal of Earth System Science, 114*(1), 37–49. https://doi.org/10.1007/BF02702007

Sanyal, T., & Chakrabarti, A. K. (1995). Farakka barrage project: Promises and achievements. *Port of Calcutta, 125*, 55–58.

Sarkar, S. K., Bhattacharya, A., Giri, S., Bhattacharya, B., Sarkar, D., Nayak, D. C., & Chattopadhaya, A. K. (2005). Spatiotemporal variation in benthic polychaetes (Annelida) and relationships with environmental variables in a tropical estuary. *Wetlands Ecology and Management, 13*(1), 55–67. https://doi.org/10.1007/s11273-003-5067-y

Sarkar, S. K., Singh, B. N., & Choudhury, A. (1986). The ecology of copepods from Hooghly estuary, West Bengal, India. *Mahasagar. Dona Paula, 19*(2), 103–112.

Sathyendranath, S. (2000). *Reports of the International Ocean-colour coordinating group* (Vol. 3, p. 140). IOCCG Project Office, IOCCG Rep.

Sharma, A., Manna, R., Satpathy, B., Naskar, M., Bhaumik, B. B., & Roshith, C. (2013). Spatio-temporal changes of hydro-chemical parameters in the estuarine part of the river Ganges under altered hydrological regime and its impact on biotic communities. *Aquatic Ecosystem Health and Management, 16*, 433–444. https://doi.org/10.1080/14634988.2013.853596

Shen, J., & Qin, Q. (2024). The general relationship between mean dissolved oxygen concentrations and timescales in estuaries. *Water, 16*(7). https://doi.org/10.3390/w16070969

Sherwill, W. S. (1858). *Report on the Rivers of Bengal: Papers of 1856, 1857, and 1858, on the Damoodah embankments*. GA Savielle, Calcutta Printing and Publishing Company (Limited).

Snoeyink, V. L., & Jenkins, D. (1991). *Water chemistry*. John Wiley & Sons.

Solanki, H. U., Dwivedi, R. M., Nayak, S., Gulati, D. K., John, M. E., & Somvanshi, V. S. (2003). Potential fishing zones (PFZ) forecast using satellite data derived biological and physical processes. *Journal of the Indian Society of Remote Sensing, 31*(2), 67–69. https://doi.org/10.1007/BF03030773

Strickland, J. D. H., & Parsons, T. R. (1972). A practical handbook of seawater analysis.

Suresh, T., Naik, P., Bandishte, M., Desa, E., Mascaranahas, A., & Matondkar, S. G. P. (2006). Secchi depth analysis using bio-optical parameters measured in the Arabian Sea. *Remote Sensing of the Marine Environment, 6406*, 408–417.

Thrush, S., Townsend, M., Hewitt, J., Davies, K., Lohrer, A., Lundquist, C., & Cartner, K. (2014). *The many uses and values of estuarine ecosystems* (pp. 226–237).

Tilley, R. J. D. (2020). *Colour and the optical properties of materials*. John Wiley & Sons.

Valerio, A. d. M., Kampel, M., Vantrepotte, V., Ward, N. D., Sawakuchi, H. O., Less, D. F. D. S., Neu, V., Cunha, A., & Richey, J. (2018). Using CDOM optical properties for estimating DOC concentrations and pCO2 in the lower Amazon River. *Optics Express, 26*(14), A657–A677. https://doi.org/10.1364/OE.26.00A657

Venkatachalapathy, R., Kumar, P., & Visweswaran, s. (2021). Mapping of Total suspended matter based on Sentinel-2 data on the Hooghly River, India. *Indian Journal of Ecology, 48*, 159–165.

Virtasalo, J., Österholm, P., & Asmala, E. (2023). Estuarine flocculation dynamics of organic carbon and metals from boreal acid sulphate soils. *Biogeosciences Discussions, 2023*, 1–32.

Vodacek, A., Hogel, F. E., Swift, R. N., Yungel, J. K., Peltzer, E. T., & Blough, N. V. (1995). The use of in situ and airborne fluorescence measurements to determine UV absorption coefficients and DOC concentrations in surface waters. *Limnology and Oceanography, 40*(2), 411–415. https://doi.org/10.4319/lo.1995.40.2.0411

Chapter 8
Chromophoric Dissolved Organic Matter (CDOM) Dynamics in Indian Coastal Waters: A Comprehensive Review

Sourav Das, Madhusmita Dash, and Chandanlal Parida

Abstract Chromophoric Dissolved Organic Matter (CDOM), the light-absorbing fraction of dissolved organic matter, plays a critical role in modulating the optical, ecological, and biogeochemical properties of coastal waters. This review comprehensively assesses CDOM dynamics along the diverse Indian coastline, spanning over 7500 km and encompassing ecosystems such as mangroves, estuaries, coral reefs, and lagoons. CDOM in Indian coastal waters arises from both allochthonous sources (e.g., riverine input, mangrove detritus) and autochthonous processes (e.g., phytoplankton exudation, microbial degradation). Its spatial-temporal variability is primarily driven by monsoonal hydrology, anthropogenic activities, and ecosystem-specific biodiversity. Notable regional differences are observed, with high CDOM levels in the Sundarbans, Kochi backwaters, and Chilika Lagoon, and relatively low concentrations in oligotrophic zones such as the Gulf of Mannar. CDOM serves as a valuable proxy for organic carbon content, nutrient enrichment, and water quality, while also influencing underwater light penetration and carbon cycling. Anthropogenic pressures—such as urban runoff, industrial discharge, and land-use changes—further alter CDOM composition and distribution. The review emphasizes the need for standardized methodologies, long-term monitoring, and region-specific remote sensing algorithms to improve our understanding of CDOM variability. It also identifies key research gaps to support coastal ecosystem management and climate-resilient policy frameworks in India.

Keywords Chromophoric Dissolved Organic Matter (CDOM) · Indian coastal ecosystems · Monsoonal variability · Remote sensing · Biogeochemical cycles · Anthropogenic impacts

---

S. Das (✉)
Department of Chemistry, Brainware University, Kolkata, West Bengal, India

M. Dash
Leibniz Institute for Baltic Sea Research, Warnemunde (IOW) Rostock, Germany

C. Parida
Curtin University, Perth, WA, Australia

© The Author(s), under exclusive license to Springer Nature Switzerland AG 2026

S. Das (ed.), *Bio-optical Properties of Indian Coastal Waters*,
https://doi.org/10.1007/978-3-032-08714-0_8

1 Introduction

Coastal waters represent some of the most dynamic and biologically productive regions on Earth, acting as vital interfaces between terrestrial and marine ecosystems. Among the various biogeochemical constituents influencing coastal processes, Chromophoric Dissolved Organic Matter (CDOM)—the light-absorbing fraction of dissolved organic matter—plays a crucial role in modulating underwater light regimes, biogeochemical cycles, and the ecological functioning of coastal and estuarine systems (Pandi et al., 2021a; Das et al., 2017; Das & Hazra, 2019).

CDOM absorbs strongly in the ultraviolet (UV) and visible blue regions of the light spectrum, significantly affecting water optical properties, such as light penetration, primary productivity, and remote sensing reflectance. It originates from both autochthonous sources, such as phytoplankton exudates and microbial byproducts, and allochthonous inputs, primarily from terrestrial runoff, decaying vegetation, and riverine discharge. Consequently, CDOM serves as a key proxy for understanding organic carbon dynamics, nutrient status, and anthropogenic influences in aquatic environments (Das et al., 2016, 2017).

The Indian coastal zone, stretching over 7500 km and encompassing a diversity of ecosystems—mangroves, estuaries, lagoons, coral reefs, and deltas—exhibits substantial spatial and temporal variability in CDOM characteristics. Factors such as monsoonal rainfall, riverine fluxes, land-use changes, and urban-industrial discharges influence CDOM dynamics along this coastline. Moreover, the unique biogeographical and climatic gradients across the east and west coasts of India further contribute to the complexity of CDOM distribution patterns (Pandi et al., 2021a; Das & Hazra, 2019).

Despite its ecological and optical significance, a comprehensive understanding of CDOM in Indian coastal waters remains limited and fragmented. Most existing studies are localized, season-specific, or focused on particular ecosystems. A synthesized understanding is essential, not only for advancing scientific knowledge but also for improving ocean color remote sensing algorithms, national carbon budget assessments, and Indian coastal water quality monitoring (Lotliker et al., 2016; Das et al., 2016).

This review aims to provide a comprehensive assessment of CDOM dynamics across major Indian coastal waters, exploring its sources, spatial-temporal variability, ecological relevance, and interactions with biodiversity and anthropogenic drivers. By integrating findings from diverse regions and methodologies, the review seeks to identify research gaps, standardize measurement practices, and highlight the importance of CDOM in the broader context of marine biogeochemistry and coastal ecosystem management.

2 Sources and Composition of CDOM

CDOM in coastal waters originates from both allochthonous and autochthonous sources.

- **Allochthonous Inputs**: Major rivers such as the Ganges, Brahmaputra, Mahanadi, Godavari, and Krishna transport substantial amounts of terrestrial organic matter into the Bay of Bengal, especially during the monsoon season. These inputs are rich in humic substances derived from soil and plant material (Pandi et al., 2014; Das et al., 2016).
- **Autochthonous Production**: Within the marine environment, phytoplankton exudation and microbial degradation contribute to CDOM levels. Studies have shown that different phytoplankton species release varying amounts and types of CDOM, influencing their optical characteristics (Pandi et al., 2021b; Das & Hazra, 2019; Das et al., 2017).

2.1 A Brief Idea on Regional CDOM Dynamics Across Indian Coastal Waters

We have found nine major coastal areas where the CDOM dynamics have been studied in the last 15 years in India (Table 8.1). These are brief ideas about-

1. Bay of Bengal (Eastern Coast).
    - Primary CDOM Sources: Major rivers like the Ganges, Brahmaputra, and Mahanadi; mangrove ecosystems.
    - Seasonal Variation: Elevated CDOM during monsoon due to increased freshwater discharge.
    - Key Influencing Factors: Monsoonal river discharge, estuarine mixing, phytoplankton activity.
    - CDOM Behavior: Conservative during dry season (15–31 PSU salinity); non-conservative during monsoon.
2. Arabian Sea (Western Coast).
    - Primary CDOM Sources: Rivers like Periyar and Narmada; urban runoff; sand mining activities.
    - Seasonal Variation: Increased CDOM in monsoon and post-monsoon periods.
    - Key Influencing Factors: Anthropogenic inputs, upwelling phenomena.
    - CDOM Behavior: High variability linked to human activities.
3. Sundarbans Estuarine System.
    - Primary CDOM Sources: Mangrove-derived organic matter; riverine inputs.
    - Seasonal Variation: Higher CDOM concentrations during monsoon.

Table 8.1 Showing the CDOM dynamics in different coastal waters of India

Region/ Coastal water	CDOM sources	Seasonal variation	Key influencing factors	CDOM behavior	References
Bay of Bengal (Eastern Coast)	Riverine input (Ganges, Brahmaputra, Mahanadi), mangroves	High during the monsoon season due to freshwater input	Monsoon discharge, estuarine mixing, phytoplankton activity	Conservative during dry season; non-conservative during the monsoon	Pandi et al. (2021a, 2021b), Das et al. (2016, 2017) and Das and Hazra (2019)
Arabian Sea (Western Coast)	River input (Periyar, Narmada), urban discharge, sand mining	Increase in monsoon and post-monsoon season	Anthropogenic influence (urbanization, sand mining), upwelling	High variability is linked to human activity	Chari et al. (2025) and Coble et al. (1998)
Sundarbans Estuarine System	Mangrove detritus, riverine organic matter	Elevated during monsoon	Mangrove export, rainfall, and tidal mixing	Seasonal peak during wet periods	Sanyal et al. (2020) and Das and Hazra (2019)
Godavari Estuary	Riverine DOM, agricultural runoff	High during the monsoon and post-monsoon season	Agricultural activities, hydrology, and phytoplankton blooms	Humic-rich CDOM during high flow periods	Chari and Pandi (2017)
Krishna Estuary	River discharge, aquaculture runoff	Peaks during the monsoon season	Freshwater inflow, nutrient enrichment	Strong land-ocean coupling	Pandi et al. (2022)
Chilika Lagoon	Wetland-derived DOM, agricultural and sewage inputs	Fluctuates with freshwater exchange	Lagoon flushing, inflow from rivers and creeks	Variable, influenced by salinity gradients	Lotliker et al. (2016), Singh et al. (2024) and Muduli and Acharya (2024)
Kochi Backwaters (Kerala)	River discharge, domestic sewage, organic detritus	Increases during southwest monsoon	Anthropogenic pressure, tidal dynamics	Localized hotspots of CDOM	Minu et al. (2014) and Prasad et al. (2024)
Gulf of Kachchh (NW India)	Tidal mixing, limited river input	Low seasonal fluctuation	High salinity, arid conditions	Generally low CDOM levels	Chari et al. (2025) and Gupta (2015); Limited studies; needs more investigation

(continued)

Table 8.1 (continued)

Region/Coastal water	CDOM sources	Seasonal variation	Key influencing factors	CDOM behavior	References
Gulf of Mannar (SE India)	Coral reef-derived organic matter, limited riverine input	Minor seasonal changes	Coral ecosystems, marine productivity	Stable but low concentration	Shaju et al. (2019) and Jyothibabu et al. (2014)

- Key Influencing Factors: Mangrove export, rainfall, tidal mixing.
- CDOM Behavior: Seasonal peaks during wet periods.

4. Godavari Estuary.

- Primary CDOM Sources: Riverine dissolved organic matter; agricultural runoff.
- Seasonal Variation: Elevated CDOM during monsoon and post-monsoon.
- Key Influencing Factors: Agricultural activities, hydrology, phytoplankton blooms.
- CDOM Behavior: Humic-rich CDOM during high flow periods.

5. Krishna Estuary.

- Primary CDOM Sources: River discharge; aquaculture runoff.
- Seasonal Variation: CDOM peaks during monsoon.
- Key Influencing Factors: Freshwater inflow, nutrient enrichment.
- CDOM Behavior: Strong land-ocean coupling.

6. Chilika Lagoon (Odisha).

- Primary CDOM Sources: Wetland-derived organic matter; agricultural and sewage inputs.
- Seasonal Variation: Fluctuates with freshwater exchange.
- Key Influencing Factors: Lagoon flushing, river, and creek inflows.
- CDOM Behavior: Variable; influenced by salinity gradients.

7. Kochi Backwaters (Kerala).

- Primary CDOM Sources: River discharge; domestic sewage; organic detritus.
- Seasonal Variation: Increases during the southwest monsoon.
- Key Influencing Factors: Anthropogenic pressure, tidal dynamics.
- CDOM Behavior: Localized hotspots of CDOM.

8. Gulf of Kachchh (Northwest India).

- Primary CDOM Sources: Tidal mixing; limited river input.
- Seasonal Variation: Low seasonal fluctuation.
- Key Influencing Factors: High salinity, arid conditions.
- CDOM Behavior: Generally low CDOM levels.

9. Gulf of Mannar (Southeast India).
   - Primary CDOM Sources: Coral reef-derived organic matter; limited riverine input.
   - Seasonal Variation: Minor seasonal changes.
   - Key Influencing Factors: Coral ecosystems, marine productivity.
   - CDOM Behavior: Stable but low concentration.

2.2 Comparison Between the Biodiversity and CDOM (Chromophoric Dissolved Organic Matter) Characteristics for Nine Major Coastal Regions of India

1. Indian Sundarbans (adjacent to Northern Bay of Bengal) (MoEF & CC, 2014)
   - Biodiversity Highlights:
     - World's largest mangrove forest, rich in estuarine species.
     - Iconic species: Royal Bengal Tiger, estuarine crocodiles, Hilsa fish.
     - Critical nursery ground for crustaceans and fishes.
   - CDOM Characteristics:
     - High CDOM (~1.2–1.5 m^{-1} at 440 nm) due to extensive mangrove detritus and riverine input from the Ganges-Brahmaputra.
     - Strong seasonal variation during monsoon due to freshwater flushing.
2. Chilika Lake (Odisha Coast) (MoEF & CC, 2014)
   - Biodiversity Highlights:
     - Asia's largest brackish lagoon with over 800 species.
     - Habitat for migratory birds, Irrawaddy dolphins, and diverse shellfish.
   - CDOM Characteristics:
     - Moderate to High CDOM (~0.8–1.2 m^{-1}) due to organic matter from wetlands, riverine input, and aquatic vegetation.
     - Significant influence of monsoon-driven river discharge.
3. Visakhapatnam Coast (Andhra Pradesh) (MoEF & CC, 2014)
   - Biodiversity Highlights:
     - Rich in plankton, pelagic fishes, and benthic invertebrates.
     - Coral patches and rocky substrates support marine biodiversity.
   - CDOM Characteristics:
     - Moderate CDOM (~0.4–0.6 m^{-1}) influenced by urban runoff and limited riverine input.

- Slight seasonal variability.

4. Chennai Coast (Tamil Nadu) (MoEF & CC, 2014)

   - Biodiversity Highlights:
     - Sandy beaches with intertidal fauna, nearshore reef patches, turtles (especially Olive Ridley nesting grounds).
     - Rich planktonic population.
   - CDOM Characteristics:
     - Moderate CDOM (~0.5–0.7 m^{-1}) mostly from urban sewage, stormwater runoff, and seasonal river discharge (Cooum and Adyar rivers).

5. Gulf of Mannar (Tamil Nadu) (MoEF & CC, 2014)

   - Biodiversity Highlights:
     - Coral reefs, seagrass beds, and over 3600 marine species, including dugongs, sea cucumbers, and ornamental fish.
     - One of the richest marine biosphere reserves in the world.
   - CDOM Characteristics:
     - Low to Moderate CDOM (~0.3–0.5 m^{-1}) due to oligotrophic offshore waters, but may increase nearshore due to fishing harbors and runoff.
     - Minimal riverine influence.

6. Kochi Backwaters (Kerala Coast) (MoEF & CC, 2014)

   - Biodiversity Highlights:
     - Rich estuarine system with mangroves, shellfish, and finfish diversity.
     - Bird habitats and nursery grounds for marine fauna.
   - CDOM Characteristics:
     - High CDOM (~1.1–1.3 m^{-1}) from organic-rich backwaters, decaying vegetation, and anthropogenic discharges.
     - CDOM increases during monsoon due to land runoff.

7. Goa Coast (Mandovi-Zuari Estuaries) (MoEF & CC, 2014)

   - Biodiversity Highlights:
     - Estuarine and coastal biodiversity including mangroves, fish, and benthos.
     - Influenced by mining, tourism, and seasonal plankton blooms.
   - CDOM Characteristics:
     - Moderate CDOM (~0.6–0.9 m^{-1}) from estuarine and mining runoff, influenced by salinity and flushing rate.
     - Seasonal peaks in pre-monsoon and monsoon.

8. Mumbai Coast (Maharashtra) (MoEF & CC, 2014)
   - Biodiversity Highlights:
     - Urban coastal ecosystem with mangroves, intertidal fauna, dolphins.
     - Coastal wetlands like Thane Creek support flamingos and birds.
   - CDOM Characteristics:
     - Moderate to High CDOM (~0.9–1.2 m^{-1}) primarily from domestic waste, industrial effluents, and degraded mangrove litter.
     - High turbidity influences optical water quality.
9. Gulf of Kachchh (Gujarat) (MoEF & CC, 2014)
   - Biodiversity Highlights:
     - Coral reefs, mudflats, mangroves, and endangered marine fauna like whale sharks and dolphins.
     - Critical region for coral regeneration.
   - CDOM Characteristics:
     - Low to Moderate CDOM (~0.4–0.6 m^{-1}) in offshore areas, higher (~0.8 m^{-1}) near estuaries and intertidal flats due to sediment-laden rivers.
     - Salinity-driven seasonal dynamics.

The Indian coastline, stretching over 7500 km, encompasses a diverse array of ecosystems, ranging from mangrove forests and estuaries to coral reefs and backwaters. This spatial heterogeneity significantly influences both the biodiversity and optical properties of coastal waters, particularly CDOM concentrations, which serve as proxies for organic carbon content, biogeochemical cycling, and water quality (Sudheesh et al., 2023; Chakraborty et al., 2020; Prasad et al., 2017; MoEF & CC, 2014).

3 Biodiversity Influence on CDOM Dynamics in the Indian Coastline

The Indian coastline supports a mosaic of diverse ecosystems, each contributing uniquely to the pool of chromophoric dissolved organic matter (CDOM). CDOM, comprising primarily humic substances, fulvic acids, and degraded phytoplankton products, is tightly linked to biological productivity and terrestrial input (Krishna et al., 2024; Vishnu et al., 2018).

In mangrove-rich areas like the Indian Sundarbans (West Bengal), Kochi (Kerala), and Mumbai's estuarine creeks (Maharashtra), CDOM levels are typically elevated due to the leaching of organic-rich litter and detritus from mangrove trees. These ecosystems act as natural carbon pumps, continuously releasing CDOM into adjacent waters through decomposition processes and tidal flushing. The dominance of refractory humic substances from mangroves often results in strong UV-absorbing CDOM, which affects light penetration and microbial activity (Das et al., 2016, 2017; Das & Hazra, 2019; Minu et al., 2014; Prasad et al., 2024).

In estuarine regions such as Mandovi-Zuari (Goa) and Chilika Lake (Odisha), the biodiversity includes fish nurseries, benthic invertebrates, seagrass beds, and migratory birds. The estuarine mixing of freshwater and seawater creates a gradient in salinity and CDOM, with peak CDOM concentrations near freshwater inflows. Seasonal phytoplankton blooms, supported by nutrient-rich waters, also contribute labile CDOM components (Menon et al., 2011; Lotliker et al., 2016; Singh et al., 2024; Muduli & Acharya, 2024).

In contrast, coral reef ecosystems like the Gulf of Mannar (Tamil Nadu) and Gulf of Kachchh (Gujarat) are generally associated with low CDOM levels in offshore waters due to their oligotrophic nature. However, localized CDOM input may arise from coral mucus, decaying seagrasses, and terrestrial runoff during the monsoon season. Excessive CDOM, especially if linked to eutrophication, can reduce water clarity, negatively impacting light-dependent coral symbionts (zooxanthellae) (Chari et al., 2025; Gupta, 2015; Shaju et al., 2019; Jyothibabu et al., 2014).

4 Seasonal and Monsoonal Changeability on CDOM Variability

The Indian monsoon system—particularly the southwest monsoon—plays a central role in modulating CDOM dynamics along the coast. During the monsoon, increased rainfall leads to high river discharge, land runoff, and terrestrial organic matter transport into coastal systems. This influx enhances CDOM concentrations significantly, particularly in river-fed systems like Sundarbans, Chilika, and Mandovi-Zuari (Das et al., 2016, 2017; Das & Hazra, 2019; Menon et al., 2011; Lotliker et al., 2016; Muduli & Acharya, 2024).

Post-monsoon, decomposition of algal blooms can also increase CDOM, but the nature of this CDOM tends to be more labile and quickly processed by microbial communities. In contrast, winter and summer months often show lower CDOM levels due to reduced runoff, increased photodegradation, and sedimentation.

This seasonal variability impacts bio-optical properties such as water color, light attenuation coefficient, and surface reflectance—parameters critical to satellite-based ocean color models (Lotliker et al., 2016).

5 Anthropogenic Impacts on CDOM Dynamics

Anthropogenic pressures such as urbanization, industrialization, agriculture, and tourism have increasingly altered CDOM sources and composition in several Indian coastal regions. For instance:

- Mumbai experiences input from untreated domestic sewage and industrial discharges, which add proteinaceous and aromatic compounds to the CDOM pool (Suresh, 2020).
- Visakhapatnam (Andhra Pradesh), an industrial hub, receives CDOM from synthetic and complex organic pollutants, altering natural CDOM optical signatures (Pandi et al., 2014).
- Chennai (Tamil Nadu) faces pressure from urban sprawl and is prone to stormwater runoff, especially during cyclonic events, which mobilize large amounts of terrestrial organic debris (Shanmugam et al., 2016).

These anthropogenic CDOM inputs are often more bioactive and can lead to hypoxia, algal blooms, and reduced biodiversity if not properly managed.

6 CDOM as a Proxy for Water Quality and Productivity

CDOM is not merely an optical interferent—it is also a useful proxy for ecosystem processes:

- It reflects carbon cycling, particularly organic matter degradation.
- It is correlated with nutrient loading, phytoplankton dynamics, and bacterial abundance.
- In estuarine and mangrove zones, CDOM is linked to carbon sequestration potential and microbial loop efficiency.

Remote sensing of CDOM using spectral absorption features is gaining prominence. However, region-specific algorithms are essential, as CDOM optical properties vary depending on source (terrestrial vs. marine), composition (humic vs. fulvic), and degradation state (fresh vs. aged).

7 Methodologies Used for CDOM Assessment in the Indian Coastal Areas

The following methods are used for CDOM characterization:

- *Spectroscopic Methods*: Absorbance and fluorescence spectroscopy are standard tools for assessing CDOM concentration and composition. Excitation-emission

matrix (EEM) spectroscopy, combined with parallel factor analysis (PARAFAC), allows for the identification of specific fluorophores (Das, 2023).
- *Remote Sensing*: Satellite-based sensors, such as Sentinel-2, facilitate large-scale monitoring of CDOM distribution and temporal changes, aiding in the assessment of anthropogenic impacts and natural variability.

8 Factors Influencing CDOM Dynamics

8.1 Hydrological Forcing

Monsoonal rainfall and river discharge are primary drivers of CDOM variability. Increased freshwater input during the monsoon transports terrestrial organic matter into coastal waters, elevating CDOM concentrations (Das et al., 2016).

8.2 Anthropogenic Activities

Land-use changes, industrial discharge, and urban runoff contribute to CDOM levels. For instance, sand mining in Kollam has been associated with higher CDOM concentrations, indicating the influence of human activities on coastal water quality (Pandi et al., 2021b).

8.3 Biological Processes

Phytoplankton blooms and microbial degradation of organic matter affect CDOM composition and concentration. Autochthonous production can lead to the formation of protein-like and humic-like substances, altering the optical properties of CDOM (Pandi et al., 2021a).

9 Implications of CDOM Variability

CDOM influences several ecological and biogeochemical processes:

- *Light Attenuation*: High CDOM levels reduce light penetration, affecting photosynthetic activity and primary productivity.
- *Carbon Cycling*: CDOM plays a role in the transport and transformation of carbon in aquatic systems, impacting carbon sequestration and emission processes.

- *Water Quality*: Elevated CDOM can indicate pollution and eutrophication, serving as a proxy for assessing the health of coastal ecosystems.

9.1 Knowledge Gaps and Future Directions

Despite advancements, several areas require further research:

- *Temporal Studies*: Long-term monitoring is needed to understand inter-annual variability and trends in CDOM dynamics.
- *Integrated Approaches*: Combining in situ measurements with remote sensing and modeling can provide comprehensive insights into CDOM behavior.
- *Anthropogenic Impact Assessment*: Quantifying the extent of human-induced changes in CDOM levels is crucial for effective management and mitigation strategies.

10 Conclusion

CDOM dynamics in Indian coastal waters are shaped by a confluence of natural processes and human activities. Understanding these dynamics is vital for managing coastal ecosystems, predicting ecological responses to environmental changes, and formulating conservation strategies. Future research should focus on integrated, multidisciplinary approaches to unravel the complexities of CDOM behavior in these diverse and dynamic environments.

Moreover, the analysis of CDOM dynamics across India's nine major coastal regions reveals a complex interplay between natural biodiversity, monsoonal hydrology, and anthropogenic stressors. High biodiversity areas like mangroves and estuaries are not only hotspots for flora and fauna but also major sources and sinks of CDOM. These areas act as natural filters, processing large volumes of organic material and supporting vital ecological services such as nutrient recycling and coastal protection.

At the same time, increasing anthropogenic influence is reshaping CDOM characteristics in urbanized coastal zones, leading to potential water quality issues. Understanding this transformation is critical in the face of climate change, sea-level rise, and coastal development, which threaten both biodiversity and water quality.

Moving forward, integrated monitoring programs combining in situ measurements, remote sensing, and biogeochemical modeling are essential. These tools can enhance our understanding of CDOM sources, fate, and transport, and can help policymakers in:

- Designing coastal zone management strategies.
- Establishing marine protected areas.
- Predicting ecosystem responses to land-use changes.

Ultimately, safeguarding India's coastal biodiversity and ensuring sustainable use of its marine resources will depend on how effectively we monitor and manage CDOM dynamics as a key environmental indicator.

References

Chakraborty, S., Ghosh, S., & Bhadury, P. (2020). Dissolved organic matter dynamics in the Indian Sundarbans: Influence of mangrove vegetation on carbon cycling. *Frontiers in Earth Science, 8*, 218. https://doi.org/10.3389/feart.2020.00218

Chari, N. V. H. K., & Pandi, S. R. (2017). Characterization of humic substances and their distribution of XAD fractions by absorption spectroscopy in the Godavari estuary, India. *Current Science, 113*, 299–303.

Chari, N. V. H., Gupta, G. V. M., Sherin, C. K., Ramu, C. V., Reddy, B. B., & Kumar, B. S. K. (2025). Variability of dissolved organic matter sources in the eastern Arabian Sea: Impacts of physical and biological processes. *Ocean Dynamics, 75*(1), 1–14.

Coble, P. G., Del Castillo, C. E., & Avril, B. (1998). Distribution and optical properties of CDOM in the Arabian Sea during the 1995 southwest monsoon. *Deep Sea Research Part II: Topical Studies in Oceanography, 45*(10–11), 2195–2223.

Das, S. (2023). Introduction to water quality science. In *An introduction to water quality science: Significance and measurement protocols* (pp. 1–17). Springer International Publishing.

Das, S., & Hazra, S. (2019). Vertical distribution of optically active substances (OAS) in the coastal sea off Indian Sundarban during post monsoon. *Marine Biology Research, 15*(2), 163–169.

Das, S., Hazra, S., & Lotliker, A. A. (2016). Delineating the relationship between chromophoric dissolved organic matter (CDOM) variability and biogeochemical parameters in a shallow continental shelf. *Egyptian Journal of Aquatic Research, 42*(3), 241–248.

Das, S., Das, I., Giri, S., Chanda, A., Maity, S., Lotliker, A. A., Kumar, T. S., Akhand, A., & Hazra, S. (2017). Chromophoric dissolved organic matter (CDOM) variability over the continental shelf of the northern bay of Bengal. *Oceanologia, 59*(3), 271–282.

Gupta, M. (2015). *Analysis of pre-and post-monsoon suspended sediments in the Gulf of Kachchh, India using remote sensing*. arXiv preprint arXiv:1503.08369.

Jyothibabu, R., Madhu, N. V., Jagadeesan, L., Anjusha, A., Mohan, A. P., Ullas, N., Sudheesh, K., & Karnan, C. (2014). Why do satellite imageries show exceptionally high chlorophyll in the Gulf of Mannar and the Palk Bay during the northeast monsoon? *Environmental Monitoring and Assessment, 186*, 7781–7792.

Krishna, A. V., Tirkey, A., Raman, M., Ali, S. M., Gupta, A., & Deb, S. K. (2024). Exploring the interplay between phytoplankton taxonomy, pigments, and optical properties in Ratnagiri coastal waters using a multi-parameter approach. *Discover Oceans, 1*(1), 23.

Lotliker, A. A., Sahoo, S., Baliarsingh, S. K., Parida, C., & Sahu, K. C. (2016). Optical characterization and assessment of ocean colour algorithms in Chilika lagoon. In *Remote sensing of the oceans and inland waters: Techniques, applications, and challenges* (Vol. 9878, pp. 205–212). SPIE.

Menon, H. B., Sangekar, N. P., Lotliker, A. A., & Vethamony, P. (2011). Dynamics of chromophoric dissolved organic matter in Mandovi and Zuari estuaries—A study through in situ and satellite data. *ISPRS Journal of Photogrammetry and Remote Sensing, 66*(4), 545–552.

Minu, P., Lotliker, A. A., Shaju, S. S., SanthoshKumar, B., Ashraf, P. M., & Meenakumari, B. (2014). Effect of optically active substances and atmospheric correction schemes on remote-sensing reflectance at a coastal site off Kochi. *International Journal of Remote Sensing, 35*(14), 5434–5447.

MoEF & CC. (2014). *Coastal and marine biodiversity of India: Conservation and sustainable utilization*. Ministry of Environment, Forests and Climate Change, Government of India.

Muduli, P. R., & Acharya, P. (2024). Monitoring of water quality of Asia's largest brackish water system, Chilika lagoon. In *Aquatic ecosystems monitoring* (pp. 65–87). CRC Press.

Pandi, S. R., Kiran, R., Sarma, N. S., Srikanth, A. S., Sarma, V. V. S. S., Krishna, M. S., Bandyopadhyay, D., Prasad, V. R., Acharyya, T., & Reddy, K. G. (2014). Contrasting phytoplankton community structure and associated light absorption characteristics of the western Bay of Bengal. *Ocean Dynamics, 64*, 89–101.

Pandi, S. R., Chari, N. V. H. K., Sarma, N. S., Chiranjeevulu, G., Kiran, R., Murthy, K. N., Venkatesh, P., Lotliker, A. A., & Tripathy, S. C. (2021a). Characteristics of conservative and non-conservative CDOM of a tropical monsoonal estuary in relation to changing biogeochemistry. *Regional Studies in Marine Science, 44*, 101721.

Pandi, S. R., Chari, N. V. H. K., Sarma, N. S., Tripathy, S. C., Chiranjeevulu, G., & Das, S. (2021b). A review of estuarine CDOM dynamics of East Coast of India influenced by hydrographical forcing. In *Estuarine biogeochemical dynamics of the East Coast of India* (pp. 223–237).

Pandi, S. R., Chari, N. V. H. K., Sarma, N. S., Lotliker, A. A., Tripathy, S. C., & Bajish, C. C. (2022). Spatiotemporal variability in the optical characteristics of dissolved organic matter in the coastal Bay of Bengal. *International Journal of Environmental Science and Technology, 19*(10), 9393–9408.

Prasad, M. B. K., Sarma, V. V. S. S., Kumar, B. S. K., & Rao, B. R. T. (2017). Variability of chromophoric dissolved organic matter (CDOM) and its absorption characteristics in the coastal waters off Visakhapatnam, India. *Journal of Marine Systems, 172*, 13–23. https://doi.org/10.1016/j.jmarsys.2017.03.002

Prasad, A., Prasad, P. R. C., & Rao, N. S. (2024). Assessing the water quality dynamics in the coastal waters of Kollam (Kerala, India) using sentinel images. *Anthropocene Coasts, 7*(1), 10.

Sanyal, P., Ray, R., Paul, M., Gupta, V. K., Acharya, A., Bakshi, S., Jana, T. K., & Mukhopadhyay, S. K. (2020). Assessing the dynamics of dissolved organic matter (DOM) in the coastal environments dominated by mangroves, Indian Sundarbans. *Frontiers in Earth Science, 8*, 1–21.

Shaju, S. S., Prasad, S. C., Vishnu, P. S., Samal, A. K., Menon, N., Nashad, M., Mathew, K. A., Nazar, A. A., & George, G. (2019). Effect of Mariculture on bio-optical properties and water quality of Gulf of Mannar and Palk Bay. *Regional Studies in Marine Science, 29*, 100618.

Shanmugam, P., Varunan, T., Jaiganesh, S. N., Sahay, A., & Chauhan, P. (2016). Optical assessment of colored dissolved organic matter and its related parameters in dynamic coastal water systems. *Estuarine, Coastal and Shelf Science, 175*, 126–145.

Singh, S., Raulo, S., Acharyya, T., Mishra, D. R., Kumar, A., Gopinath, A., Muduli, P. R., & Rastogi, G. (2024). Dynamics and source characterization of chromophoric dissolved organic matter (CDOM) in a tropical monsoon-driven lagoon. *Marine Chemistry, 265*, 104430.

Sudheesh, V., Suresh, T., Sudharma, K. V., & Vethamony, P. (2023). Coupling between chromophoric dissolved organic matter and dissolved inorganic carbon in Indian estuaries: Implications to estuarine acidification. *Science of the Total Environment, 893*, 164810. https://doi.org/10.1016/j.scitotenv.2023.164810

Suresh, T. (2020). *Characterization of optical properties of Mandovi and Zuari estuarine waters*. Doctoral dissertation, Goa University.

Vishnu, P. S., Shaju, S. S., Tiwari, S. P., Menon, N., Nashad, M., et al. (2018). Seasonal variability in bio-optical properties along the coastal waters off Cochin. *International Journal of Applied Earth Observation and Geoinformation, 66*, 184–195.

Chapter 9
Advancing Ocean Colour Remote Sensing in India: Capacity Building Initiative

Sanjiba Kumar Baliarsingh, Susmita Raulo, Alakes Samanta, Aneesh A. Lotliker, T. V. S. Udaya Bhaskar, Sudheer Joseph, and T. M. Balakrishnan Nair

Abstract Ocean Colour Remote Sensing (OCRS) and marine bio-optics play a crucial role in understanding marine ecosystems, biogeochemical cycles, and climate change. Given the significance of the Indian Ocean and its teleconnections in hydro-biological and climate-related processes, India has made substantial advancements in recent decades in OCRS. These include the successful launch of ocean colour missions (Oceansat series) and the development of expertise and infrastructure to implement OCRS technology for oceanographic and climate studies. This comprehensive review emphasizes the importance and applications of OCRS in India while also highlighting capacity building efforts to enhance research and practical applications in this field. The International Training Centre for Operational Oceanography, functioning under the Indian National Centre for Ocean Information Services, has played a key role in organizing annual short-term training programs in OCRS over the past decade, benefiting more than 2500 participants, primarily from India and the Indian Ocean rim countries. These programs primarily emphasize data availability, the use of specialized software, data analysis, and environmental monitoring, equipping trainees with the skills to incorporate OCRS into their research endeavours. This review highlights the critical need for such training initiatives to empower participants in utilizing OCRS for the development of ocean value chain services. India has an advanced academic and research framework for OCRS, with several institutions and universities (~20 numbers) offering specialized courses or incorporating OCRS-related topics into undergraduate and postgraduate programs, covering its fundamentals and applications. Significant contributions from international organizations/initiatives such as POGO, IOCCG, and PORSEC have also been instrumental in OCRS capacity building in India. Key challenges and future prospects of OCRS in India include the development of region-specific algorithms, fostering multinational and interdisciplinary collaborations, and enhancing relevant infra-

S. K. Baliarsingh (✉) · S. Raulo · A. Samanta · A. A. Lotliker · T. V. S. Udaya Bhaskar · S. Joseph · T. M. B. Nair
Indian National Centre for Ocean Information Services, Ministry of Earth Sciences, Government of India, Hyderabad, India

© The Author(s), under exclusive license to Springer Nature Switzerland AG 2026

S. Das (ed.), *Bio-optical Properties of Indian Coastal Waters*,
https://doi.org/10.1007/978-3-032-08714-0_9

structure. To address these challenges, it is essential to introduce intensive, long-duration training programs and integrate dedicated courses into Indian academia, particularly at the postgraduate level in oceanography and environmental studies.

Keywords Ocean colour · Remote sensing · Training · INCOIS · Oceansat

1 Introduction

Ocean Colour Remote Sensing (OCRS) is a powerful tool for monitoring biological productivity, algal dynamics, water quality, carbon cycle, and many more (Fig. 9.1). In general, OCRS measures the optical properties of sea water that are determined by the interaction of solar irradiance with optically active substances such as suspended matter, chlorophyll, and coloured dissolved organic matter (IOCCG, 2000;

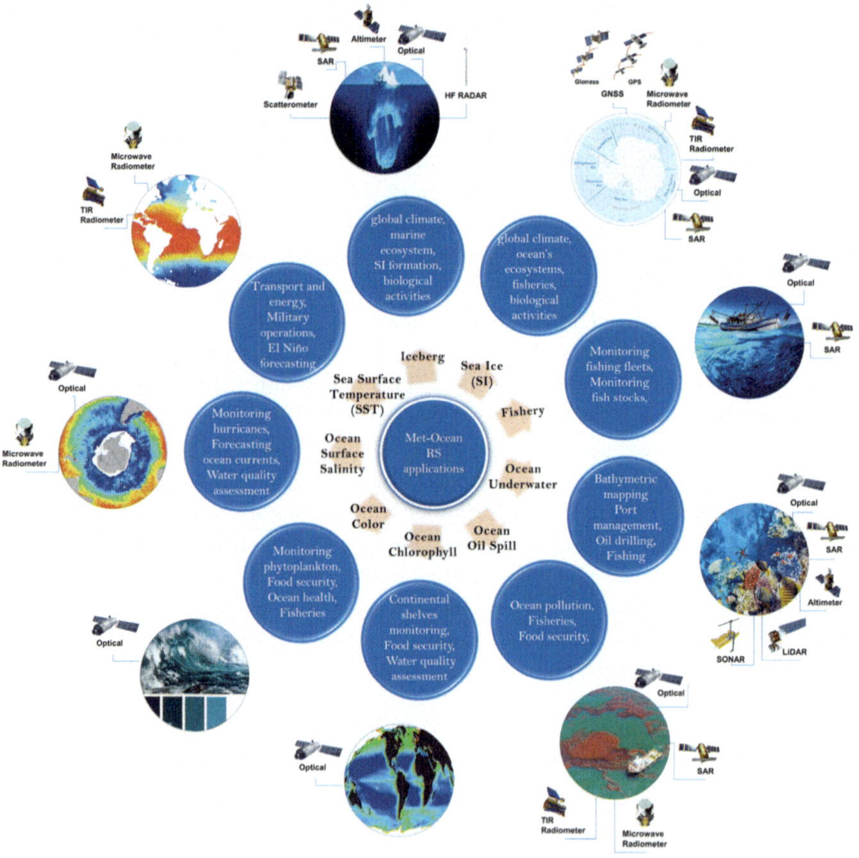

Fig. 9.1 Overview of the met-ocean applications of remote sensing. (Source: Amani et al., 2022)

Behrenfeld et al., 2005; Gordon & Morel, 2012; Gholizadeh et al., 2016). Marine bio-optics complements OCRS by exploring these optical interactions at various depths that allow precise interpretation of remote sensing data. OCRS and marine bio-optics are essential for monitoring marine ecosystems, primary productivity, and biogeochemical processes in the ocean. These techniques help to study phytoplankton dynamics, sediment transport, harmful algal blooms, and climate change impacts.

Globally, space agencies conduct OCRS by launching and maintaining satellite missions. Notable agencies and their key missions include the United States' National Aeronautics and Space Administration (SeaWiFS, MODIS, PACE), the European Space Agency (Sentinel series, EnMAP), the United States' National Oceanic and Atmospheric Administration (VIIRS, GOES-R), France's Centre National d'Études Spatiales (POLDER, SWOT), the Korea Aerospace Research Institute (GOCI), and the Indian Space Research Organization (Oceansat series).

The field of OCRS has undergone significant advancements over the past few decades. This progress began with the launch of NASA's Coastal Zone Colour Scanner (CZCS) aboard the Nimbus-7 satellite in 1978, marking the first global-scale ocean colour observations (Gordon et al., 1980; Hovis et al., 1980) and offering an unprecedented view of the ocean's biological productivity (McClain, 2009). Subsequent missions, including SeaWiFS (1997), MODIS-Aqua (2002), MERIS (2002), and VIIRS (2011), enhanced spectral resolution, spatial coverage, and temporal frequency of ocean colour data (Hooker et al., 1992; Donlon et al., 2012). More recently, advanced hyperspectral missions such as PACE, Sentinel-3, and GOCI have provided high-resolution data, further improving ocean studies (Gregg & Casey, 2004; Werdell et al., 2019). These technological advancements have facilitated precise monitoring of ocean colour, biogeochemical cycles, and ecosystem dynamics at both global and regional scales.

Effectively interpreting and utilizing ocean colour data demands specialized expertise in bio-optical modeling, algorithm development and refinement, sensor calibration, and data assimilation techniques. This underscores the need for capacity building in the field. In this regard, infrastructure development and international collaboration have played a crucial role in enhancing capabilities in OCRS. Various global institutions dedicated to OCRS research have undertaken initiatives to support capacity building through specialized training programs, workshops, and conferences. Notable contributors include NASA's Applied Remote Sensing Training Program (ARSET) program, the International Ocean Colour Coordinating Group (IOCCG) summer lecture series, European Organisation for the Exploitation of Meteorological Satellites (EUMETSAT) training sessions, and regional training programs organized by Partnership for Observation of the Global Ocean (POGO), among others.

Over the past few decades, India has made remarkable advancements in OCRS through dedicated satellite missions such as the Oceansat series, research initiatives, and both national and international collaborations. Indian academic institutions and research organizations have played a crucial role in developing algorithms, refining methodologies to enhance satellite data accuracy, and creating application-driven models. India has emerged as a global leader in OCRS research, with significant contributions from organizations like the Indian Space Research Organisation (ISRO) and the Indian National Centre for Ocean Information Services (INCOIS).

Recognizing the importance of capacity building, India has undertaken various research initiatives, training programs, and academic curriculum developments to strengthen expertise in OCRS and marine bio-optics. INCOIS, in particular, has been instrumental in conducting capacity building programs by integrating training sessions, institutional collaborations, technology development, and international partnerships. These programs provide training in satellite oceanography, hands-on experience with bio-optical parameter measurements, OCRS data processing, applications, and algorithm development.

With the growing need for monitoring marine ecosystems, particularly for water quality assessment and climate studies, enhancing capacity in this field becomes crucial for advancing both national and global scientific progress. Against this backdrop, a comprehensive review has been conducted to emphasize the significance and application of OCRS in India. This review specifically highlights capacity building initiatives aimed at strengthening research and practical applications in this domain. The primary objective is to assess India's contributions to capacity building in OCRS, ultimately supporting efforts in marine ecosystem monitoring.

2 India's Ocean Colour Remote Sensing Journey

India's OCRS program has undergone significant evolution, initially relying on sensors such as CZCS, SeaWiFS, MODIS, and MERIS (Desa et al., 2001; Solanlki et al., 2001; Chaturvedi et al., 1998). These datasets played a crucial role in building expertise, ultimately leading to the development of India's own OCRS satellites. The adoption of OCRS in India facilitated a wide range of ocean-related applications, including coastal zone studies, under a major initiative called "Marine Satellite Information Services" (Sridhara Murthy et al., 1999). This program was designed to develop methodologies for retrieving various oceanographic parameters and utilizing them for key application areas. A notable outcome of these efforts is the routine advisories on potential fishing zones based on satellite-derived sea surface temperature and chlorophyll gradients, significantly aiding the fishing community. Additionally, coastal zone mapping at various scales has been carried out using IRS satellite data, contributing to planning, conservation, and developmental strategies. Advanced methodologies have also been established for retrieving oceanographic parameters such as ocean surface waves, winds, wave forecast modeling, internal waves, oceanic eddies, heat budgets, mixed layer depth, and latent heat studies. These advancements have been made possible using satellite data from various international missions.

Many oceanographic phenomena, including coastal dynamics and climate change, exhibit variations over timescales ranging from hours to years. The development of operational applications for these phenomena necessitates continuous, long-term satellite data availability. Recognizing this need, along with the promising results from experimental ocean studies and the increasing meteorological interest in sea state dynamics, India initiated a mission to develop and launch indigenous satellites for ocean applications. The first in a planned series of ocean-focused satellites was IRS-P4, also known as Oceansat-1, launched in early 1999. It carried an Ocean Colour Monitor (OCM), specifically designed to provide regional and global observations for measuring biological constituents, such as chlorophyll pigments, and assessing primary productivity, catering to the major needs of the user community. The following section briefs OCM and its follow-ups.

3 Indian Ocean Colour Mission: Oceansat Series

India's journey into OCRS studies began with the launch of IRS-P4 (Oceansat-1) in 1999, which carried the OCM sensor (Sridhara Murthy et al., 1999). This milestone marked a significant step in India's oceanographic research, providing critical data on ocean productivity, water quality, and primary productivity. The satellite played a vital role in fisheries management, offering insights into potential fishing zones and marine ecosystem health (Groom et al., 2019).

Building on this success, Oceansat-2, launched in 2009, featured an improved OCM with higher resolution and enhanced OCRS capabilities (Nagamani et al., 2008). Unlike its predecessor, Oceansat-2 carried a comprehensive set of oceanographic payloads, including a scatterometer, altimeter, thermal infrared radiometer, and microwave radiometer. This mission provided a full suite of ocean observations, including wind fields, wave spectrum, sea surface topography, ocean currents, internal waves, sea surface temperature, and chlorophyll concentration. The satellite played a crucial role in detecting algal blooms, improving climate models, and refining operational oceanographic services. A major focus of the program was to enhance the accuracy of ocean parameter retrieval, assimilate oceanographic data into meteorological models, and develop innovative applications for ocean and atmospheric studies.

Advancing further, Oceansat-3, launched in 2022, introduced next-generation sensors with significantly improved spatial and spectral resolution (Sankhala et al., 2025). Equipped with an advanced OCM-3 sensor, a scatterometer (SCAT-3), and a Sea Surface Temperature Monitor (SSTM), this satellite delivers high-precision data for marine ecosystem studies, atmospheric research, and oceanographic applications. Its refined capabilities support coastal resource management, environmental monitoring, climate change studies, and disaster preparedness. By offering enhanced insights into phytoplankton dynamics, ocean pollution, and atmospheric aerosols, Oceansat-3 has strengthened India's ability to monitor and predict oceanographic changes, contributing to sustainable marine resource management and global climate research.

4 Key Applications of Ocean Colour Remote Sensing in India

4.1 Marine Ecosystem Monitoring

The OCRS has the capability to measure several geophysical parameters that provide information on optically active substances (e.g., chlorophyll, suspended matter, dissolved organic matter) which are crucial in understanding the ecosystem ecology (Robinson, 1983; McClain, 2009; Sathyendranath et al., 2017; Groom et al., 2019; Darecki & Stramski, 2004). Measurements of these features and their variations over space and time can be interpreted to provide information relating to biological productivity, fish location, water quality, sedimentation patterns, pollution, and other phenomena (Platt & Sathyendranath, 2008; Krug et al., 2017). Some of the major applications of OCRS include coastal zone management, measurement of phytoplankton and assessment of their distribution, detection of algal blooms and their dynamics, estimating ocean productivity, identifying potential fishing zones, monitoring coral reefs, estimation of optical properties and phytoplankton abundance for marine resource and habitat assessment, understanding the marine carbon cycle, observation of pollution and sediment inputs to the coastal zone and their impact on marine food, etc. (Amani et al., 2022).

4.2 Operational Products and Services: INCOIS' Pivotal Role

The growing understanding of OCRS applications in monitoring marine ecosystems and coastal environments has reinforced its role as a powerful tool for societal benefit. With increasing environmental challenges such as climate change, ocean acidification, and marine pollution, OCRS-based data provides real-time insights essential for sustainable ocean resource management and conservation efforts.

In India, the INCOIS plays a pivotal role in developing and disseminating OCRS-based near real-time products and advisory services supporting a broad spectrum of applications, including water quality assessment, fisheries management, climate studies, and monitoring coastal resources. INCOIS has operationalized several key products, such as the Marine Fisheries Advisory Service (MFAS), Algal Bloom Information Service (ABIS), and Coral Bleaching Alert System (CBAS), while additional products are under development to further expand OCRS applications (Baliarsingh et al., 2024).

Under MFAS, INCOIS provides Potential Fishing Zone (PFZ) advisories, along with species-specific advisories for Tuna and Hilsa fisheries. These advisories are generated by analyzing conducive environmental conditions and habitat suitability, primarily using OCRS-derived data, enabling fishermen to optimize their catch while promoting sustainable fishing practices. The ABIS utilizes chlorophyll concentration data to track phytoplankton blooms, offering valuable insights into ocean productivity, fisheries management, and early warning systems for harmful algal blooms that may impact marine biodiversity and human health. The CBAS is instrumental in monitoring coral reef health by detecting temperature-induced bleaching events. Tracking sea surface temperature anomalies aids in conservation efforts, climate resilience strategies, and ecosystem management, ensuring proactive measures can be taken to safeguard coral reefs against environmental stressors. With these advancements, India's OCRS-based services continue to enhance marine resource management, environmental protection, and climate adaptation strategies, reinforcing the nation's commitment to sustainable ocean governance and societal well-being.

5 Ocean Colour Capacity Building Activities: Indian Context

The effective application of OCRS in tackling critical marine and environmental challenges requires a highly skilled workforce capable of processing and interpreting complex satellite-derived data. Therefore, capacity building is crucial to unlocking the full potential of OCRS data. To achieve this, various initiatives such as training programs, workshops, conferences, and academic courses play a pivotal role in equipping scientists, researchers, policymakers, and resource managers with the necessary expertise to utilize OCRS products efficiently. In India, several research institutions and universities offer specialized training programs, academic degrees, and short-term courses designed to enhance proficiency in OCRS data processing and application. Additionally, collaborations with international organizations further enhance India's capabilities by facilitating knowledge exchange and adopting cutting-edge tools and methodologies. Through strategic investments in capacity building, India continues to cultivate expertise in the sustainable utilization of OCRS data, thereby contributing to effective marine resource management and environmental conservation.

5.1 By International Training Centre for Operational Oceanography

Capacity building and specialized training, particularly in developing nations, remain an ongoing necessity to fully harness the potential of OCRS resources and datasets for the global user community. Recognizing the critical role of capacity development in ocean science and technology, the INCOIS and UNESCO's Intergovernmental Oceanographic Commission (IOC-UNESCO) formalized their commitment by signing a Memorandum of Agreement on July 4, 2013, during the 27th session of the IOC Assembly in Paris. This agreement facilitates collaboration in conducting training programs at the International Training Centre for Operational Oceanography (ITCOocean), established by INCOIS in Hyderabad, India. ITCOocean operates as a key contributor to IOC-UNESCO's global training and capacity development initiatives. It closely collaborates with the International Oceanographic Data and Information Exchange (IODE) program and serves as a regional training hub under the Ocean Teacher Global Academy (OTGA). The center is dedicated to advancing the scientific foundation, technological innovation, and information systems necessary for operational oceanography at national, regional, and global scales.

The primary objective of ITCOocean is to provide specialized training in operational oceanography to young scientists, technical experts, and policymakers from India, Indian Ocean Rim nations, and other developing countries. By offering continuous and structured training, ITCOocean aims to cultivate a highly skilled workforce equipped with expertise in oceanographic data processing and interpretation, fostering a robust knowledge base within the subcontinent.

In the domain of OCRS and bio-optics, ITCOocean has been actively conducting training programs on OCRS data processing, analysis, and applications in collaboration with global organizations such as OTGA, the IOCCG, and the POGO. These initiatives have significantly contributed to the worldwide capacity building efforts in OCRS and bio-optics, empowering researchers and professionals with cutting-edge skills and tools for ocean monitoring and management. A detailed account ITCOocean's contributions to capacity development in OCRS and bio-optics is presented in Table 9.1.

The OCRS and related training programs offered by ITCOocean are primarily conducted in a hybrid mode, integrating both virtual sessions and hands-on field campaigns for the collection and analysis of bio-optical parameters. These programs are designed to provide comprehensive training in key areas, including the fundamentals of OCRS, bio-optical parameter processing and analysis, algorithm development and optimization for OCRS applications, and the utilization of satellite-derived data products for ecosystem service development, with a particular focus on INCOIS'services.

ITCOocean has welcomed participants from India and various countries worldwide, particularly from the Indian Ocean Rim nations, further cementing INCOIS' recognition as a leading institution in capacity building for OCRS (Fig. 9.2). The majority of these training programs have been conducted in collaboration with leading international organizations such as the OTGA, the IOCCG, and the POGO. Over time, the frequency and scale of OCRS training initiatives have steadily increased, driven by consistently positive feedback from participants. This enthusiastic response has further encouraged ITCOocean to expand its offerings, reinforcing its commitment to nurturing the next generation of OCRS experts and advancing global expertise in OCRS.

Table 9.1 Details of ocean colour (and related) capacity building training programs carried out by the International Training Center for Operational Oceanography (ITCOocean), India

Sl. no	Year	Duration	Title	Number of participants Total	India	Other countries
1	2014	24–29 March	Remote Sensing of Potential Fishing Zones and Ocean State Forecast	24	15	9
2	2014	10–14 November	International Training Course On: Ocean Colour Remote Sensing—Data, Processing and Applications	28	20	8
3	2017	6–10 March	OTGA/IODE-IOC Training Course on "Discovery and Use of Operational Ocean Data Products and Services"	18	11	7
4	2018	19–23 March	Training programme on "Remote Sensing of Marine Phytoplankton—optics, pigment and taxonomy"	20	20	0
5	2018	18–22 June	OTGA-INCOIS Training Course On "Discovery and Use of Operational Ocean Data Products and Services"	31	17	14
6	2018	18–19 September	Training Course on "Operational Ocean Services, Data and Data Products"	Indian coast guard officers		
7	2019	25–29 March	Practical course on "Marine Phytoplankton—Optics, Pigment and Taxonomy"	18	18	0
8	2019	01–05 July	OTGA-INCOIS Training Course On "Discovery and Use of Operational Ocean Data Products and Services" during	32	20	12
9	2019	25–29 November	Ocean teacher global academy Training course on: Ocean color remote sensing—Data, processing and analysis	37	24	13
10	2020	31 August–04 September	Online Training Course on "Discovery and Use of Operational Ocean Data Products and Services"	102	80	22
11	2020	16–20 November	Online Course on Fishery Oceanography for Future Professionals (Level: Basic, Batch-1)	1380	903	477
12	2021	10–23 April	Online Course on Fishery Oceanography for Future Professionals (Level: Basic, Batch-2)			
13	2021	26–30 July	Principles of Ocean Remote Sensing and Its Applications	328	153	175
14	2021	25 October–05 November	OTGA-INCOIS Training course on "Discovery and use of Operational Ocean data products and services"	56	13	43

(continued)

Table 9.1 (continued)

Sl. no	Year	Duration	Title	Number of participants		
				Total	India	Other countries
15	2021	08–12 November	Biological Observations in the Indian Ocean (From Microbes to Megafauna) Organized by Indian Ocean Global Ocean Observing System (IOGOOS)	70	24	46
16	2022	05–09 September	"Fundamentals of Remote Sensing & Its Oceanographic Applications"	206		
17	2022	31 October–05 November	POGO—ITCOocean Training Program on "Ocean Observations to Societal Applications"	23	15	8
18	2022	12–16 December	OTGA-INCOIS Training Course: "Ocean Color Remote Sensing", Data Processing & Analysis	10	06	04
19	2023	20–24 February	"Training Programme on Ocean Observation System and Models" towards OSF, Coastal Dynamics and PFZin joint collaboration with Andhra University and National Institute of Oceanography -RC, Visakhapatnam	43	43	00
20	2023	09–10 March	Training Course on "Ocean Observation System, operational Services and Ocean Data Utilization" for students from IIT Bhubaneswar and University of Hyderabad	17	17	00
21	2023	10–14 April	"Fundamentals of Remote Sensing & GIS and Oceanographic Applications" for students of Marine Science from Berhampur University	50	50	00
22	2023	07–11 August	"Oceanographic Remote Sensing: Bridging the Gap between Fundamentals and Applications"	246		
23	2023–24	October 2023–January 2024	Short-term course spanning 4–6 months for Official from SNOM, Indian Navy			
24	2023	04–08 December	OTGA-INCOIS Training Course: "Ocean Color Remote Sensing—Data, Processing and Analysis"	36	28	08
25	2024	18 January–07 February	ITEC—ITCOocean Training Program on Fishery Oceanography for the Ocean Decade (F.O.O.D.)—2024			

(continued)

Table 9.1 (continued)

Sl. no	Year	Duration	Title	Number of participants		
				Total	India	Other countries
26	2024	29 January–07 February	POGO—ITCOocean Training Program on "Ocean Observations for Coastal Applications" in collaboration with CEMACS (Malaysia), SUST (Bangladesh), SQU (Oman) and Andhra Uni., Visakhapatnam	30	11	19
27	2024	08–12 April	"Fundamentals of Remote Sensing & GIS and Oceanographic Applications" for students of Marine Science from Berhampur University	36	36	00
28	2024	21–25 October	OTGA-INCOIS Training Course: "Ocean Color Remote Sensing—Data, Processing and Analysis"	21	17	04

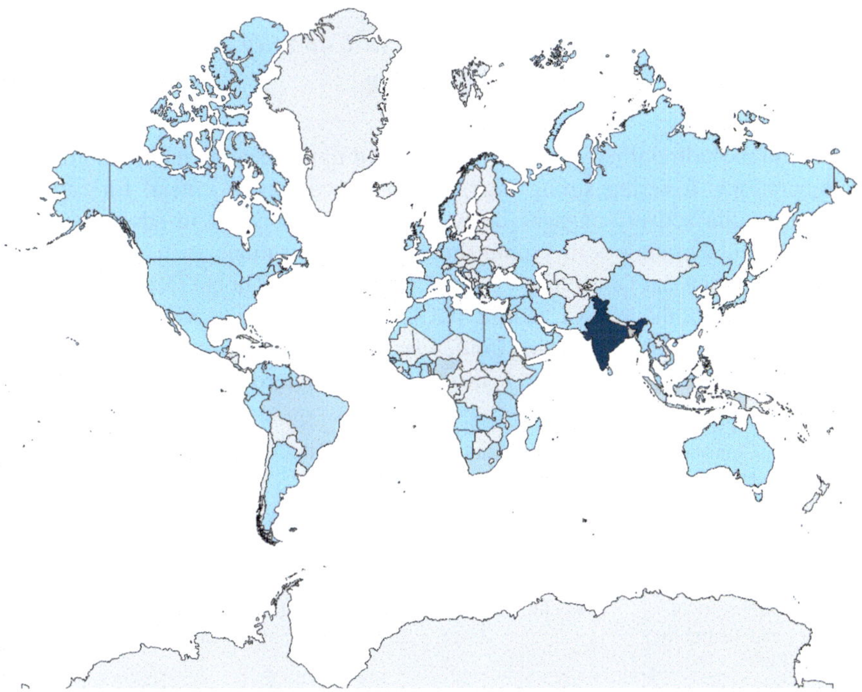

Fig. 9.2 A detailed account of the contribution of the International Training Center for Operational Oceanography, India, in the global ocean colour capacity building training programs. The countries of the participants are highlighted in different shades of blue

5.2 Sporadic Capacity Building Programs by Other Indian Institutions

Alongside INCOIS, other institutions such as the National Remote Sensing Centre (NRSC, ISRO), Space Applications Centre (SAC, ISRO), and National Institute of Oceanography (NIO, CSIR), Cochin University of Science and Technology (CUSAT), and Indian Society of Remote Sensing (ISRS) have also contributed to OCRS capacity building by offering specialized courses, training modules, and research opportunities. Among them, NRSC provides regular training on remote sensing and GIS applications, whereas other institutions have conducted the training occasionally. ISRS conducts workshops and conferences on all aspects of remote sensing, including OCRS for monitoring marine ecosystems. These events provide a platform for professionals, researchers, and students to gain a basic understanding and knowledge of advanced methodology through the exchange of ideas and knowledge (Table 9.2).

5.3 International Collaborations w.r.t Capacity Building in Ocean Colour Remote Sensing

Given the multidisciplinary nature of OCRS and its reliance on multi-institutional collaborations, fostering strong international networking is crucial for capacity building. India actively engages in international partnerships to advance OCRS research, collaborating with prominent global organizations such as the Indian Ocean Global Ocean Observing System (IOGOOS), the Global Ocean Observing System (GOOS), the POGO, and the IOCCG. Additionally, it works closely with the Pacific Ocean Remote Sensing Conference (PORSEC), the International Society for Photogrammetry and Remote Sensing (ISPRS), Chlorogin (Chlorophyll Global

Table 9.2 Contribution of Indian institutions, other than INCOIS, to the capacity building programs related to ocean colour remote sensing and bio-optics

Institution	Location	Trainings/workshops
National Remote Sensing Centre (NRSC), ISRO	Hyderabad, Telangana	Remote Sensing & GIS Application
Cochin University of Science and Technology (CUSAT)	Kochi, Kerala	Training Programme on Remote Sensing of Ocean Colour
Andhra University and National Institute of Oceanography -RC, Visakhapatnam	Hyderabad, Telangana, Visakhapatnam, Andhra Pradesh	Training Programme on Ocean Observation System and Models towards OSF, Coastal Dynamics and PFZ in joint collaboration with INCOIS
Indian Society of Remote Sensing (ISRS)	All India	Annual Symposia and Workshops on remote sensing technology in various fields

Integrated Network), Societal Application in Fisheries and Aquaculture using Remotely Sensed Imagery (SAFARI), the Trevor Platt Science Foundation (TPSF), and EUMETSAT, among others. The Indian institutions such as National Institute of Oceanography (NIO), Space Applications Centre (SAC), National Remote Sensing Agency (NRSA), Centre for Marine Living Resources and Ecology (CMLRE), Indian National Centre for Ocean Information Services (INCOIS), are the active Indian partners, in conducting specific trainings, workshops, and conferences in collaboration with the aforementioned international organizations. These strategic alliances not only enhance India's research expertise but also contribute to global scientific initiatives, fostering knowledge exchange and creating new opportunities for innovation and development (Table 9.3). Among all Indian partners, INCOIS is dedicated toward conducting courses on OCRS either as a part of other trainings or specific trainings on OCRS data processing and application. All the programs focused on overall applications of OCRS specifically in the field of fisheries and coastal applications, due to their importance in sustainable management of marine resources. The trainings and workshops aimed in exerting technical expertise to create new generation of ocean experts, whereas the conferences allows to adopt appropriate and advanced technologies in the field of OCRS.

5.4 Ocean Colour in the Curriculum of Indian Academia

India has a well-established academic and research ecosystem for OCRS, with several institutions offering specialized programs integrated into their curricula (Table 9.4). Several universities, such as Cochin University of Science and Technology (CUSAT), Goa University, Annamalai University, Andhra University, Berhampur University, and Kerala University of Fisheries and Ocean Studies (KUFOS), provide master's programs in marine sciences, marine biology, oceanography, satellite oceanography, and remote sensing. These programs include specific courses, units, or chapters that focus on the principles and applications of OCRS.

Furthermore, premier technological institutions like IIT Madras and IIT Kharagpur offer M.Tech. programs in ocean engineering, ocean technology, and geoinformatics, emphasizing the role of OCRS in marine and coastal studies. The Indian Institute of Remote Sensing (IIRS) also plays a pivotal role in capacity building, providing specialized degree programs and short-term courses related to OCRS application in environmental monitoring. Through the concerted efforts of these universities and research institutions, India continues to nurture a highly skilled workforce in marine sciences. These institutions not only drive advancements in marine research and coastal resource management but also contribute significantly to climate change studies, reinforcing India's expertise and leadership in the field of OCRS.

Table 9.3 International collaborations w.r.t. capacity building

Training/workshops	Indian partnership	International partnership
Interdisciplinary multi-sensor studies of the Pacific Ocean	National Institute of Oceanography (NIO), Goa	Pan Ocean Remote Sensing Conference (PORSEC) Association
International Training Course on Ocean Colour: Techniques and Applications during February 12–23, 2001	Space Applications Centre (SAC), Indian Space Research Organisation (ISRO)	International Ocean Colour Coordinating Group (IOCCG) and Partnership for Observation of Global Ocean (POGO)
Ninth IOCCG Committee Meeting during 15–17 January 2004, Hyderabad, India	National Remote Sensing Agency (NRSA), Indian Space Research Organisation (ISRO)	International Ocean Colour Coordinating Group (IOCCG)
Remote Sensing Applications in Fisheries and Ecosystem Analysis during 11–13 February 2010, Kochi, India	National Institute of Oceanography, Kochi	SAFARI (Societal Application in Fisheries & Aquaculture using Remotely Sensed Imagery, AG-06-02)
International Symposium on Remote Sensing and Fisheries during 15–17 February 2010, Kochi, India	National Institute of Oceanography, Kochi	SAFARI (Societal Application in Fisheries & Aquaculture using Remotely Sensed Imagery, AG-06-02)
Ocean Remote Sensing for Well-Being of All, during 5–9 November 2012 in Kochi	Centre for Marine Living Resources and Ecology (CMLRE), Kochi Indian National Centre for Ocean Information Services (INCOIS), Hyderabad	Pan Ocean Remote Sensing Conference (PORSEC) Association
International Training Course: Ocean Colour Remote Sensing: Data, Processing and Applications during 10–14 Nov 2014, Hyderabad, India	International Training Centre for Operational Oceanography (ITCOcean) ESSO-INCOIS	IODE OceanTeacher Global Academy (OTGA)
Biological Observations in the Indian Ocean (From Microbes to Megafauna) during 08–12 Nov 2021	Indian National Centre for Ocean Information Services (INCOIS)	Indian Ocean Global Ocean Observing System (IOGOOS)
POGO—ITCOocean Training Program on “Ocean Observations for Coastal Applications” during 29 Jan–07 Feb 2024	Indian National Centre for Ocean Information Services (INCOIS) and Andhra University, Visakhapatnam	POGO Centre for Marine and Coastal Studies (CEMACS), Malaysia Shahjalal University of Science and Technology (SUST), Bangladesh Sultan Qaboos University (SQU) Oman
2024 IOCCG Summer Lecture Series on Frontiers in Ocean Optics and Ocean Colour Science during November 4–16, 2024	Indian National Centre for Ocean Information Services (INCOIS)	Ocean Carbon & Biogeochemistry (OCB) Project Office, Woods Hole, MA, USA EUMETSAT (European Organisation for the Exploitation of Meteorological Satellites) International Ocean Colour Coordinating Group (IOCCG)

Table 9.4 Details of the courses offered by Indian institutions and universities that include ocean colour remote sensing as a part of the academic curriculum

Institution	Location	Program(s) offered
Cochin University of Science and Technology (CUSAT)	Kochi, Kerala	M.Sc. in Marine Biology
Andhra University	Visakhapatnam, Andhra Pradesh	M.Sc. in Oceanography
Berhampur University	Berhampur, Odisha	M.Sc. in Marine Biology and M.Sc. in Oceanography
Kerala University of Fisheries and Ocean Studies (KUFOS)	Kochi, Kerala	M.Sc. in Oceanography and Ocean Modeling
Indian Institute of Technology (IIT) Madras	Chennai, Tamil Nadu	M.Tech. in Ocean Engineering
Indian Institute of Technology (IIT) Kharagpur	Kharagpur, West Bengal	M.Tech. in Ocean Engineering and Naval Architecture
Indian Institute of Remote Sensing (IIRS)	Dehradun, Uttarakhand	M.Tech. in RS&GIS with specialization "Marine and Atmospheric Sciences" NNRMS short Course (for University/Institute/college Teachers) Summer school on "Usefulness of RS&GIS for Environmental Science"
Pondicherry University	Puducherry	M.Sc. in Marine Biology
Alagappa University	Karaikudi, Tamil Nadu	M.Sc. in Oceanography and Coastal Area Studies
Goa University	Taleigao Plateau, Goa	M.Sc. in Marine Sciences
Bharathidasan University	Tiruchirappalli, Tamil Nadu	M.Sc. in Marine Science
Annamalai University	Chidambaram, Tamil Nadu	M.Sc. in Marine Biology
Madurai Kamaraj University	Madurai, Tamil Nadu	M.Sc. in Marine Biology
Amity University	Noida, Uttar Pradesh	M.Sc. in Marine Science
Marine Engineering and Research Institute (MERI)	Mumbai, Maharashtra	B.Sc. in Maritime Science
University of Mumbai	Mumbai, Maharashtra	M.Sc. in Oceanography
Dr. B.R. Ambedkar University	Delhi, India	M.Sc. in Environmental Sciences with a focus on Oceanography
Vels University	Chennai, Tamil Nadu	B.Sc. in Maritime Studies M.Sc. in Marine Science
Institute of Ocean Management, Anna University	Chennai, Tamil Nadu	M.Tech. in Ocean Technology Ph.D. in Ocean Management
University of Kerala	Thiruvananthapuram, Kerala	M.Sc. in Oceanography Ph.D. in Marine Sciences
Jawaharlal Nehru Technological University (JNTU)	Hyderabad, Telangana	M.Tech. in Geoinformatics with focus on ocean colour monitoring

5.5 *Ocean Colour Remote Sensing and Marine Bio-Optics Research Programs in India*

Research and development programs play a crucial role in enhancing capacity building efforts within their respective fields through human resource development, academic degrees, and research publications. While organizations such as the Space Applications Centre (SAC) and the National Remote Sensing Centre (NRSC) of ISRO have conducted several validation programs for Oceansat series, available records indicate that India had one specific dedicated research initiative on OCRS and marine bio-optics, known as the "Satellite Coastal and Oceanographic Research (SATCORE)" program, executed by INCOIS during 2008–2017.

Satellite-based retrieval of ocean colour parameters requires an accurate bio-optical algorithm, particularly in coastal waters, which in turn necessitates a comprehensive in situ database for a specific region. To address these challenges, the SATCORE program was implemented with the primary objective of validating satellite-derived ocean geophysical parameters in coastal waters and facilitating their application in operational services, such as PFZ identification and algal bloom detection and monitoring. Additionally, SATCORE aimed to develop new algorithms for OCRS satellite sensors, refine atmospheric correction techniques, and utilize OCRS data for modeling primary productivity. INCOIS 'SATCORE program conducted long-term measurements of bio-optical properties at 11 time-series stations along the Indian coastline in collaboration with various research institutions and universities (Fig. 9.3).

Key achievements of the SATCORE program include the establishment of the first extensive in situ bio-optical parameter database for Indian coastal waters, validation of OCRS products, and development of regional bio-optical algorithms. The program also led to the development of algorithms for remote estimation of phytoplankton types, the Algal Bloom Index, and bloom monitoring. Additionally, a novel aerosol correction algorithm was developed for precise ocean radiance retrieval in algal bloom-dominated waters, and critical inputs were provided to ISRO for defining visible channel-based bands for the OCM-3 sensor onboard Oceansat-3. Through these achievements, SATCORE significantly contributed to capacity building in OCRS by training numerous research scholars and students, marking it as a pioneering and successful research initiative in the field.

6 Summary and the Way Forward

This comprehensive review presents an in-depth assessment of India's progress in OCRS and marine bio-optics capacity building initiatives, highlighting their critical role in monitoring marine ecosystems, assessing climate change impacts, and understanding biogeochemical cycles and operational services. Over the years, India has made remarkable progress in OCRS through dedicated satellite missions

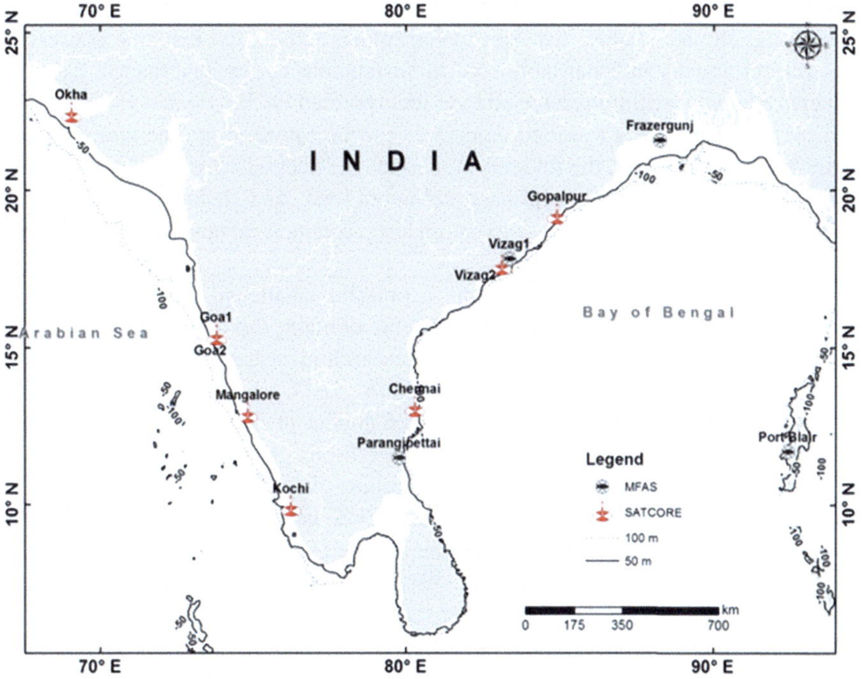

Fig. 9.3 Map showing time-series stations sustained under the Indian National Centre for Ocean Information Services (INCOIS)' Satellite Coastal and Oceanographic Research (SATCORE) program in the Indian coastal waters

such as the Oceansat series and sustained capacity building efforts, significantly enhancing research capabilities and practical applications in marine sciences.

A focal point of this review is India's structured training programs, particularly those spearheaded by the ITCOocean at INCOIS. These initiatives have been instrumental in equipping researchers, policymakers, and students with expertise in OCRS through specialized training workshops, skill-development programs, and curriculum integration in higher education. Furthermore, international collaborations with organizations such as POGO, IOCCG, and PORSEC have played a pivotal role in strengthening India's capacity building framework, fostering knowledge exchange, and promoting global best practices in OCRS research.

Despite these achievements, certain challenges persist in India's OCRS development. Key issues include the need for region-specific bio-optical algorithms for Indian coastal waters, the expansion of international collaborations, and improvements in research infrastructure. To address these gaps, this review advocates for the expansion of training programs, the integration of dedicated OCRS courses in postgraduate education, and the promotion of interdisciplinary research initiatives. Notably, the SATCORE program has been a milestone effort, contributing significantly to the validation of satellite-derived oceanographic parameters and the development of regional OCRS algorithms.

This review underscores the diverse applications of OCRS in marine ecosystem monitoring, climate studies, fisheries management, and coastal resource conservation, reinforcing its indispensable role in sustainable ocean governance. Several Indian academic institutions have already incorporated OCRS-related subjects into their curricula, fostering a skilled workforce for the future of marine research and technology. Additionally, the review emphasizes the necessity for long-term investments in research, cross-disciplinary collaborations, and deeper integration of OCRS within academic institutions to ensure sustained advancements in ocean monitoring.

However, capacity building remains a pressing challenge for India's OCRS research community. To further strengthen this domain, the review highlights the urgent need for developing state-of-the-art research infrastructure, including dedicated laboratories, high-performance computing (HPC) facilities, and advanced data storage systems. Additionally, the establishment of dedicated oceanographic cruise programs for hands-on training, enhanced inter-institutional collaborations for experiential learning, and specialized training programs focusing on AI and machine learning applications for real-time OCRS data processing and predictive modeling are recommended as key strategies for the future. To further enhance long-term engagement and knowledge sharing within the OCRS community, the review recommends the creation and sustained operation of an online common forum-cum-consortium. This platform would serve as a hub for capacity building initiatives, facilitating seamless interaction among trainees, researchers, and experts, thereby nurturing continuous learning and innovation in the field in India.

Acknowledgements The authors express gratitude to the Director, INCOIS, for spearheading the ocean colour remote sensing activities and capacity building initiatives and for encouragement to carry out this study. This is INCOIS' publication contribution number 570.

Data Availability All the information utilized in this study has been gathered from publicly available information on various institutional websites, given the nature of this review. The event dates and locations provided by these sources are included in this compilation.

References

Amani, M., Moghimi, A., Mirmazloumi, S.M., Ranjgar, B., Ghorbanian, A., Ojaghi, S., Ebrahimy, H., Naboureh, A., Nazari, M.E., Mahdavi, S., Moghaddam, S.H.A., Asiyabi, R.M., Ahmadi, S.A., Mehravar, S., Mohseni, F., & Jin, S. (2022). Ocean remote sensing techniques and applications: A review (Part I). *Water, 14*(21), 3400.

Amani, M., Mehravar, S., Asiyabi, R. M., Moghimi, A., Ghorbanian, A., Ahmadi, S. A., Ebrahimy, H., Moghaddam, S. H. A., Naboureh, A., Ranjgar, B., & Mohseni, F. (2022). Ocean remote sensing techniques and applications: A review (Part II). *Water, 14*(21), 3401.

Baliarsingh, S.K., Samanta, A., Lal, D.M., Mohanty, P.C., Raulo, S., Giri, S., Dash, B., Jha, S., Sathish, C., Premkumar, R., Joseph, S., Nair, T.M.B., & Kumar, T.S. (2024). Advancing marine ecological services for the Indian Seas: INCOIS's contribution. *Indian Journal of Geo-Marine Sciences, 53*(05), 276–293.

Behrenfeld, M. J., Boss, E., Siegel, D. A., & Shea, D. M. (2005). Carbon-based ocean productivity and phytoplankton physiology from space. *Global Biogeochemical Cycles, 19*(1).

Chaturvedi, N., Narain, A., & Pandey, P. C. (1998). Remote sensing applications in oceanography with special reference to bio-physical coupling: A review. *Indian Journal of Marine Science, 27*, 1–9.

Darecki, M., & Stramski, D. (2004). An evaluation of MODIS and SeaWiFS bio-optical algorithms in the Baltic Sea. *Remote Sensing of Environment, 89*(3), 326–350.

Desa, E. S., Suresh, T., Matondkar, S. P., & Desa, E. (2001). *Sea truth validation of sea WiFS Ocean colour sensor in the coastal waters of the eastern Arabian Sea*. Indian Academy of Sciences.

Donlon, C. J., et al. (2012). The global monitoring for environment and security (GMES) Sentinel-3 mission. *Remote Sensing of Environment, 120*, 37–57.

Gholizadeh, M. H., Melesse, A. M., & Reddi, L. (2016). A comprehensive review on water quality parameters estimation using remote sensing techniques. *Sensors, 16*(8), 1298.

Gordon, H. R., & Morel, A. Y. (2012). *Remote assessment of ocean color for interpretation of satellite visible imagery: A review*. Wiley.

Gordon, H. R., Clark, D. K., Mueller, J. L., & Hovis, W. A. (1980). Phytoplankton pigments from the Nimbus-7 coastal zone color scanner—comparisons with surface measurements. *Science, 210*, 63–66.

Gregg, W. W., & Casey, N. W. (2004). Global and regional evaluation of the SeaWiFS chlorophyll data set. *Remote Sensing of Environment, 93*(4), 463–479.

Groom, S., Sathyendranath, S., Ban, Y., Bernard, S., Brewin, R., Brotas, V., Brockmann, C., Chauhan, P., Choi, J. K., Chuprin, A., & Ciavatta, S. (2019). Satellite Ocean colour: Current status and future perspective. *Frontiers in Marine Science, 6*, 485.

Hooker, S. B., Brown, J. W., & Balch, W. M. (1992). *SeaWiFS technical report series*. NASA.

Hovis, W. A., Clark, D. K., Anderson, F., Austin, R. W., Wilson, W. H., Baker, E. T., et al. (1980). NIMBUS-7 coastal zone color scanner—System description and initial imagery. *Science, 210*, 60–63.

International Ocean Colour Coordinating Group [IOCCG]. (2000). *Remote sensing of ocean color in coastal and other optically-complex waters*. Bedford Institute of Oceanography.

Krug, L. A., Platt, T., Sathyendranath, S., & Barbosa, A. B. (2017). Ocean surface partitioning strategies using ocean colour remote sensing: A review. *Progress in Oceanography, 155*, 41–53.

McClain, C. R. (2009). A decade of Satellite Ocean color observations. *Annual Review of Marine Science, 1*, 19–42.

Nagamani, P. V., Chauhan, P., & Dwivedi, R. M. (2008). Development of chlorophyll-$ a $ algorithm for ocean colour monitor onboard OCEANSAT-2 satellite. *IEEE Geoscience and Remote Sensing Letters, 5*(3), 527–531.

Platt, T., & Sathyendranath, S. (2008). Ecological indicators for the pelagic zone of the ocean from remote sensing. *Remote Sensing of Environment, 112*(8), 3426–3436.

Robinson, I. S. (1983). Satellite observations of ocean colour. *Philosophical Transactions of the Royal Society of London. Series A, Mathematical and Physical Sciences, 309*(1508), 415–432.

Sankhala, D. K., Rani, S. I., Srinivas, D., Prasad, V. S., & George, J. P. (2025). Validation of OceanSat-3 sea surface winds for their utilization in the NCMRWF NWP assimilation systems. *Advances in Space Research, 75*(2), 1945–1959.

Sathyendranath, S., Brewin, R. J., Jackson, T., Mélin, F., & Platt, T. (2017). Ocean-colour products for climate-change studies: What are their ideal characteristics? *Remote Sensing of Environment, 203*, 125–138.

Solanlki, H. U., Dwivedi, R. M., Nayak, S. R., Jadeja, J. V., Thakar, D. B., Dave, H. B., & Patel, M. I. (2001). Application of ocean colour monitor chlorophyll and AVHRR SST for fishery forecast: Preliminary validation results off Gujarat coast, northwest coast of India. *Indian Journal of Geo-Marine Sciences, 30*(3).

Sridhara Murthy, K. R., Gowda, H. H., & Desai, N. M. (1999, February). Future Ocean satellites & their applications: Indian perspectives. In *Cooperation in space* (Vol. 430, p. 329).

Werdell, P. J., et al. (2019). The plankton, aerosol, cloud, ocean ecosystem mission: Status, science, advances. *Bulletin of the American Meteorological Society, 100*(9), 1775–1794.

Index

© The Editor(s) (if applicable) and The Author(s), under exclusive license to Springer Nature Switzerland AG 2026
S. Das (ed.), *Bio-optical Properties of Indian Coastal Waters*,
https://doi.org/10.1007/978-3-032-08714-0

The manufacturer's authorised representative in the EU is Springer Nature Customer Service Centre GmbH, Europaplatz 3, 69115 Heidelberg, Germany. If you have any concerns regarding our products, please contact ProductSafety@springernature.com

Printed and bound by CPI Group (UK) Ltd, Croydon, CR0 4YY
05/10/2026
02228044-0003